CRITICAL CHOICES AND CRITICAL CARE

Philosophy and Medicine 51

CATHOLIC STUDIES IN BIOETHICS 2

CRITICAL CHOICES AND CRITICAL CARE

Catholic Perspectives on Allocating Resources in Intensive Care Medicine

Edited by

KEVIN Wm. WILDES, S.J.

Assistant Professor of Philosophy, Senior Research Scholar of the Kennedy Institute of Ethics and Center for Clinical Bioethics, Georgetown University, Washington D.C., U.S.A.

KLUWER ACADEMIC PUBLISHERS

DORDRECHT / BOSTON / LONDON

Library of Congress Cataloging in Publication Data

Critical choices and critical care : Catholic perspectives on allocating resources in intensive care medicine / edited by Kevin Wm. Wildes.
p. cm. -- (Philosophy and medicine ; v. 51. Catholic studies in bioethics ; v. 2)
Includes index.
ISBN 0-7923-3382-9 (hbk. : acid-free paper)
1. Critical care medicine--Religious aspects--Catholic Church. I. Wildes, Kevin Wm. (Kevin William), 1954- . II. Series: Philosophy and medicine ; v. 51. III. Series: Philosophy and medicine. Catholic studies in bioethics ; v. 2.
RC86.95.C75 1995
241'.6424'08822--dc20 95-1907

ISBN 0-7923-3382-9

Published by Kluwer Academic Publishers,
P.O. Box 17, 3300 AA Dordrecht, The Netherlands

Kluwer Academic Publishers incorporates the publishing programmes of D. Reidel, Martinus Nijhoff, Dr. W. Junk and MTP Press.

Sold and distributed in the U.S.A. and Canada by Kluwer Academic Publishers, 101 Philip Drive, Norwell, MA 02061, U.S.A.

In all other countries, sold and distributed by Kluwer Academic Publishers Group, P.O. Box 322, 3300 AA Dordrecht, The Netherlands.

Printed on acid-free paper

Printed in the Netherlands

TABLE OF CONTENTS

PART THREE: APPENDICES

PREFACE

The genesis of this book was an on-going conversation between Catholic and non-Catholic physicians, philosophers, and theologians. However, substantive conversations rarely "just happen". They need support and attention. The Steering Committee on Bioethics of the International Federation of Catholic Universities, with generous foundational support, has given this conversation the attention and support it so needed. Begun in 1981 the Steering Committee, under the leadership of Francesc Abel, S.J. and John Collins Harvey has fostered numerous conversations for Catholic thinkers on issues in bioethics. This particular conversation, on the moral questions of intensive care medicine, could not have happened without their leadership. They were supported in their work by Lisa Cahill, L.B.J. Stuyt, and Robert Cefalo. H. T. Engelhardt, Jr. was invaluable to the conversation in both his leadership in its planning and his substantive participation. Francesc Abel, S.J. was vital to the project in his participation and in providing a place – an atmosphere – in which the group could meet, discuss, argue, and explore ideas.

The final form of this conversation on moral issues in critical care medicine has taken shape in this book. I am grateful to the authors for their work in preparing their manuscripts and for their patience with me. A special note of thanks is due to Angela Jacobson and Charlene Brown for their help and patience in preparing this manuscript.

The International Federation of Catholic Universities has reorganized the Bioethics Group and made it part of the Association of Catholic Medical Schools. This change insures that there will be continuing interdisciplinary work and reflection by Catholic scholars and institutions in the area of bioethics. It is our hope that these reflections will continue to develop as a body of literature that nourishes practitioners, theologians, and philosophers throughout the world.

Kevin Wm. Wildes, S.J.
Georgetown University
Washington, D.C.

KEVIN WM. WILDES, S.J.

THE MORAL PARADOX OF CRITICAL CARE MEDICINE

Contemporary medicine is practiced in a paradox. On the one hand medicine offers seemingly infinite possibilities for sustaining and improving human existence. Debilitating diseases can be identified and cured. Organs can be transplanted and the lives of a recipients transformed. Childless couples can now have children. Each day genetic studies increase our understanding of diseases and offer the hope of eventual cure. With the development of medical support for major organ systems it seems that death can be postponed indefinitely. Modern medical practice appears to have no boundaries. Yet, on the other hand, the practice of medicine is constrained by limits. There are limits to our lives. We will all die. There are limits to our knowledge. We do not always know why a patient is ill, how the patient can be cured, or why a patient becomes better. And there are limits to our resources. We do not have the resources, in personnel, time, or money, to do everything that can be done for each patient. Each decision to do good for a patient is an implicit choice not to do something for someone else.

As we become aware of the limits of our resources we can understand the relationship between choices made in the clinic and the impact on social policy. The choices made in the clinic not only affect the patient who is treated but such decisions also affect other, anonymous, patients who will be treated less aggressively or, perhaps, not at all. At the same time the choices made in the arena of social policy influence the range of choices that are possible in the clinic. The paradox then is that while medicine seems to offer infinite possibilities the practice of medicine is governed by limits.

No other area of medicine highlights this paradox more clearly than critical care medicine. One has only to consider the three following realities. First, patients admitted to Intensive Care Units tend to be among those who are sickest and in need of the most help. They are often confronting their own deaths. Second, to give these patients the type of support they need intensive care units require vast resources. For example, it is estimated that in the United States one percent (1%) of the

K. Wm. Wildes (ed.), Critical Choices and Critical Care, 1–4.

Gross Domestic Product is spent annually on intensive care units. The decision to dedicate resources to such units is also an implicit decision to limit resources in other areas of medicine. Finally, like other areas of medicine, critical care confronts us with the limits of our medical knowledge. Critical care medicine reminds us that our knowledge is statistically based. Choices are made without the comfort of certitude. While intensive care medicine can support life infinitely it is practiced in the context of limits.

This paradox of medicine's possibilities and limits is the context for difficult moral questions. Faced with the possibility and hope to do good we must decide how to act when we cannot do everything that is good. Critical care medicine provides a point of focus for such moral reflection. The moral dilemmas confronted in critical care medicine are posed by our finitude. That is, we are limited creatures with limited resources. We must make choices among competing goods. The confrontation with finitude is a deeply Christian theme. Christians understand themselves as creatures dependent on a loving Creator. This Creator has given men and women life and the gifts of creation. These lives and resources are ours to steward. With an understanding of finitude and stewardship Christianity brings a rich tradition of spiritual and moral reflection to bear on the moral questions in critical care medicine.

The themes and questions of finitude have been a central part of Roman Catholic moral theology (Cronin). In both theological reflection and pastoral practice the Church has confronted the obligation to do good in a world where not all goods can be done. Catholic physicians, patients, and their families have struggled with these moral quandaries in the practice of medicine while theologians, chaplains, and bishops have sought to understand and articulate the obligation to prolong life and the obligation to provide health care for all men and women.

This book is an attempt to bring together the moral questions of intensive care medicine with Roman Catholic reflections on the obligations to prolong life and to provide health care for all men and women. The conclusions that can be learned from critical care medicine will provide helpful insights in other areas of medical practice and the formulation of health policy. This book comes from an on-going conversation among its authors with the hope of inviting others into the conversation. Physicians, theologians, and philosophers have taken up the moral questions posed by the practice of critical care medicine with the hope of addressing the questions of the clinic as well as those of social policy.

Critical care medicine has opened new horizons for both healing and dying. The techniques and practice of critical care medicine have allowed surgeons to explore procedures that would not be possible without the recovery support made possible by intensive care. The essays in the first section of this book explore the historical development and practice of critical care medicine and the moral questions it raises. These essays set out the background for the moral questions surrounding intensive care medicine and decisions at the end of life. As Baruch Brody's essay makes clear the advancement of many medical techniques can be tied to the development and use of critical care medicine. Intensive care medicine provides necessary support for other areas of medical intervention (e.g. cardiac surgery). However, because of the vast resources required in the practice of critical care medicine there are moral questions about its use. Brody looks at questions about admission, discharge, and the role of the physician. Who should be admitted to such units? Who should be excluded? What are the criteria that should guide us in such decisions? How much should a society devote to critical care medicine? The essays by Herman Delooz, Pedro Rosso and Alejandro Serani-Merlo illustrate how the practice of critical care medicine, and its moral problems, transcend national boundaries and have influenced the practice of medicine globally. Michael Rie argues in his essay that the use of critical care medicine is a social question as well as a clinical question. Since critical care medicine consumes great amounts of resources it also leads to other questions about how we are deploying the limited resources at our disposal.

The essays in the second section of the book bring different aspects of Roman Catholic Christian theology to bear on understanding and resolving these questions. Since the medical advances of the sixteenth century the Roman Catholic Christian community has struggled with questions, in medicine, about the obligation of patients to purse treatment. Within its moral framework Christianity has understood human life as a basic good to be preserved. While the proscription against taking human life is clear in Christian thought, the question that emerged in the sixteenth century concerned the extent to which one must pursue medical treatment in order to preserve life. If there is an obligation to preserve life, must that obligation be pursued at all costs? In developing the distinction between ordinary and extraordinary means the Catholic community has sought to come to terms with these questions. The essays by Kevin Wildes and Diego Gracia set out the historical and systematic development of this tradition. The essay by Paul Schotsmans moves beyond the traditional framework of ordinary and extraordinary means to a theological framework that examines the relational aspects of human life as a way of articulating these questions of critical care in the clinic and in public policy.

The moral questions of critical care medicine are not restricted to the treatment of particular patients as the essays in the first part of the book point out. Again, the Catholic community has a tradition of speaking to issues of public policy and social questions. The essays by Cathleen Kaveny and Tom Kopfensteiner attempt to bring the social teaching of the Church together with its medical moral tradition in confronting the questions of critical care medicine.

The moral dilemmas and controversies raised by the social perspective are at least three. First, there is a commitment to see that all men and women have basic health care. This is a theme that has been part of the Church's social teaching in the last century. Second, there are questions about the stewardship of resources that must be asked. The use of common resources is a moral question not just a question of economy or efficiency. Choices must be made about the goods pursued with common resources. Health care is a complex good created by both public and private resources. It is a good in which people benefit differently. The third moral issue concerns how health policy is created. A society must ask what are the moral parameters for creating health policy. Should it be created by a group of "experts" or is there a moral, as well as political, obligation to create such policy in the public forum with open participation?

This book has developed with the hope that it will lead to further discussion of the issues of how health care resources are deployed. It is created to bring discussion between physicians and health policy analysts into dialogue with theologians and philosophers. It is conceived with the hope of bringing two elements of the Catholic tradition, the social teaching and the medical-moral tradition, into discussion with one another. Nevertheless it is a limited, finite project designed to address moral questions in one area of contemporary medical practice in the light of the insights of theological reflections. In so doing the book sets the stage for future discussions.

REFERENCE

Cronin, D.A.: 1989, *Conserving Life*, Pope John XXIII Medical-Moral Research and Educational Center, Braintree, MA.

PART ONE

CLINICAL QUESTIONS

BARUCH A. BRODY

ETHICAL ISSUES RAISED BY INTENSIVE CARE

Over the last twenty years, considerable attention has been paid to ethical issues arising out of the provision of care in intensive care units (ICUs). This is not surprising. After all, so much of the literature of medical ethics is an attempt to evaluate the appropriateness of medical interventions (usually high-technology interventions) designed to prolong life when the effectiveness in terms of life-extension is uncertain, the resulting quality of the life prolonged is questionable, and the costs (economic and otherwise) of providing the intervention are high. This type of evaluation is particularly needed in intensive care units where, because of the severity of the patient's illness, the outcomes are especially uncertain and questionable and the costs are extremely high.

This essay will not attempt to develop an ethical response to these many issues. It will, instead, aim at identifying and clarifying the crucial issues. After an opening section which presents some historical, descriptive, and demographic information about intensive care units, the rest of the paper will be divided into two major sections. One section will address the many complex issues surrounding the criteria for admission to and discharge from intensive care units when the focus is just on the question of whether or not this is beneficial to the patient. The other section will address the further complexities that arise when economic considerations play a role, either because they have resulted in fewer facilities than claimants so a selection of who will and who will not receive intensive care is inevitable (the triaging context) or because families and/or society are requesting that costs be controlled so that a selection of who will and who will not receive intensive care is desired even though it is not inevitable (the cost-containment context). In such cases, the physician is serving as a gatekeeper to the intensive care unit, and the further complexities reflect the moral ambiguities involved in this gatekeeping role.

K. Wm. Wildes (ed.), Critical Choices and Critical Care, 7–22.

I. A PRIMER OF INTENSIVE CARE

The history of intensive care units has been sketched elsewhere (Bryan-Brown, 1992), so we can just briefly summarize it here. One source of the modern intensive care unit was the surgical recovery room which slowly evolved into the surgical intensive care unit (SICU). As early as the 1860s, Florence Nightingale had advocated the use of special areas with skilled nurses to carefully monitor post-surgical patients. Johns Hopkins had a special neurosurgical postoperative unit in the 1920s. The increasing demands of ambitious cardiac surgery gave rise to special units in the 1950s, beginning both in Boston and in San Francisco. Eventually, this concept spread to most hospitals. A second source of the modern intensive care unit was the experience of the Blegdam Hospital in Copenhagen when it confronted the terrible polio epidemic of 1952. Needing to manage an unprecedented number of patients in respiratory failure, it set up a special unit (using 250 medical students and 260 extra nurses) who ventilated patients with manual intermittent positive-pressure ventilation. Despite the sense of crisis in which everyone worked, fundamental principles such as control of the airway by a tracheostomy and the regular clearing of secretions from the lungs were established. These ideas spread throughout Europe and then to the United States. Danish expatriates helped set up a major medical intensive care unit(MICU) at the Massachusetts General Hospital, and a similar unit was set up by Peter Safar and his colleagues, first at Johns Hopkins and then in Pittsburgh. Such units became particularly important with the development of cardiopulmonary resuscitation in the early 1960s. Still a third source of the modern intensive care unit was the discovery by cardiologists in the 1960s that since the majority of deaths in patients who had a myocardial infarction and who had been admitted alive to the hospital were from treatable dysrhythmias, careful cardiac monitoring followed by early aggressive treatment of dysrhythmias could lead to significant drops in mortality rates. Special units, often called coronary care units(CCU), were developed to provide that monitoring and early aggressive treatment. The last source of the modern intensive care unit was the development of the neonatal intensive care unit(NICU) as a location for the management of premature infants. Ever since Tarnier in 1878 introduced the first "warming chamber" (incubator), it has been recognized that one can increase the survival rate of premature infants by better specialized management. With the application of aggressive respiratory support to these premature infants, the modern neonatal intensive care unit emerged.

This historical survey also provides us with a useful introduction to the different medical functions of an intensive care unit. Some of these relate to the monitoring of the functioning of major systems, especially the cardio-pulmonary system. This type of monitoring is provided in the form of continual (often computerized) ECG monitoring, by pulse oximetry monitoring, by apnea monitoring, by invasive arterial monitoring, *etc.* Others relate to the provision of support for failing systems, including medications (to correct dysrhythmias and strengthen the functioning of failing hearts, to treat widespread infection, etc.), mechanical ventilatory support, transvenous pacemakers, hemodialysis, and artificial nutrition/hydration. In thinking about the ethical issues raised by the use of intensive care units, it is extremely important to differentiate both these different types of functioning and these different classes of patients. The provision of certain types of intensive care for certain types of patients may have a very different moral status than the provision of other types of intensive care for other patients.

We are fortunate in having two important surveys (OTA, 1984 and Groeger, 1992) of adult intensive care units in the United States; I know of no comparable data from other countries. The earlier survey showed that adult intensive care units comprised 5.9% of total hospital beds in short-term non-federal hospitals, with an additional 0.9% of the beds being devoted to pediatric and neonatal intensive care units. This represented a major growth in the late 1970s and early 1980s, perhaps stimulated by more generous Medicare reimbursements for intensive care unit beds, since there were no obvious technical improvements during this period which were sufficient to explain this growth. The costs for patients in these 5.9% of beds were estimated to run about 3 times as much as costs for other patients, taking into account all hospital charges but excluding separately billed physician charges. This represents the greater cost of the extra manpower (especially nursing, where the nurse-patient ratio ranges from 1:1 to 1:3) involved in intensive care unit care, of the technology used, and of the ancillary services (e.g., lab work). While difficult to estimate, it was thought that 20–30% of the patients were in the intensive care units for monitoring, and were receiving only routine physician services. This percentage was much higher in coronary care units. At the other extreme were the patients whose care involved constant direct physician involvement and the use of advanced life-support equipment. These usually represented less than half of the patients, and that percentage dropped to the 10–20% range in some units. In between these two extremes were the 30–70% of patients who were receiving one or more stabilizing interventions but whose care did not require constant physician involvement. Finally, the report emphasized that there was insufficient information about outcomes to properly assess the effectiveness of ICU care, but that there were reasons to be skeptical about whether it really helped the extremely sick patient who had a severe chronic underlying illness (e.g., cancer or cirrhosis) and about whether it was really needed in many of the patients who were just being monitored. The latter survey showed that the percentage of intensive care units had continued to increase, now comprising more than 8% of all beds. It also showed that as hospitals got larger, they tended to have more specialized units, rather than a single intensive care unit. Finally, it provided valuable information about how these units were run. Decision making was very diffused in most units. For example, admission to the intensive care unit was controlled by the director or by an intensive care unit attending physician in only 12% of all units (with the highest percentage–30%-being in pediatric and neonatal units). Any physician with hospital admitting privileges could admit a patient to an intensive care unit in the majority of hospitals. Even houseofficers had such admitting privileges in many hospitals, especially in the larger teaching hospitals. There was a similar diffusion of authority in the writing of medical orders and in the discharge process. The significance of this diffusion of authority for our ethical issues will emerge below.

Several equally useful surveys exist for neonatal intensive care units in the United States (OTA, 1981; OTA, 1987; Hack, 1991). All remind us that the majority of patients in neonatal intensive care units are there because of a very low birth weight associated with prematurity. This needs to be kept in mind, because it means that any measures (e.g., better prenatal care) which can reduce the incidence of prematurity will also reduce the tremendous social and personal costs (monetary and otherwise) associated with neonatal intensive care. Nevertheless, this point can be exaggerated. Recent data (Hack, 1991) indicates that 80% of the women delivering very low birthweight babies had at least some prenatal care.

Between 150,000 and 200,000 infants in the United States were admitted to neonatal intensive care units (OTA 1981, 1987). The length of hospital stay (for survivors) varied from a mean of 114 days in the very smallest infants (501–750 grams), to 80 days in the next smallest group (751–1000 grams), to 46 days in a somewhat larger group (1001–1500 grams) (Hack, 1991). Not surprisingly, these smallest babies, who stay in the hospital for a very long time, are among the most expensive admissions to any hospital.

What are the outcomes produced by this tremendous use of neonatal intensive care? For a variety of reasons, this question has been extensively studied, especially for the low birthweight infant, and we probably know more about the outcome of neonatal intensive care for that class of patients than about any other form of intensive care for any other class of patients. In order to understand this outcome data, one needs to keep in mind that we are concerned not merely with survival rates but also with the incidence of major morbidities. Among these major morbidities are several which are unfortunately associated with the very ventilatory support these children need to survive. These include chronic lung disease (bronchopulmonary dysplasia), blindness (retinopathy of prematurity), and neurological damage due to bleeding in the brain (intraventricular hemorrhages).In the middle of the 1980s, the good news was the major improvements in outcomes in the 751–1000 gram infant(OTA, 1987). The most recent data (Hack, 1991) shows major improvements in even smaller infants. Survival rates averaged 53% in the 700–800 gram infant, 33% in the 600–700 gram infant, and 21% in the 500–600 gram infant. Even more exciting is the fact that almost half the survivors in this group survived without significant morbidities. This data, it must be stressed, is from the period just before the extensive use of antenatal steroids and administering surfactant to newborns, two interventions that are expected to produce further improvements in neonatal outcomes in the 1990s.

We have tried in this section to produce an overall picture of intensive care. Perhaps the most crucial theme that has emerged from our brief survey is that intensive care is a very diverse phenomenon, and that any good ethical analysis of intensive care will have to take that diversity into account. This theme will be even more prevalent in the next two sections of this paper.

II. ICU PROGNOSIS AND MEDICALLY BENEFICIAL CARE

Recent discussions in medical ethics (President's Commission, 1983; Hastings Center, 1987) seem to have produced widespread agreement on the following claims about the foregoing of life-sustaining treatment, whether in an intensive care unit or elsewhere: (1) The mere fact that some treatment exists which may prolong the patient's life for some period of time is not by itself sufficient reason to provide that treatment. In that way, the technological imperative is not a moral imperative; (2) Life-prolonging therapy may be withheld and/or withdrawn when it is judged that the burdens of the treatment are sufficiently great that they cannot be justified by the possible benefits of life-prolonging therapy; (3) When that point is reached is a value decision, rather than a medical decision, and that decision must therefore be made by the patient (or those, the surrogates, who speak for the patient) in light of the patient's values; (4) The role of the physician in that decisional process is to accurately assess the patient's condition and the patient's prognosis under the available treatment options (including palliative care) and to provide that information to the patient or the patient's surrogates.

There are many detailed questions that need to be raised about this consensus. One can ask about claim (1) whether there are some modalities of therapy (e.g., food and fluids) that must always be provided. One can ask about claim (2) whether the withholding/withdrawing is equally valid if it is the resulting life, rather than the therapies, which is judged burdensome (this is the double effect-issue). One can ask about claim (3) whether surrogates are supposed to make substituted judgements or best-interest judgements. But I want to focus on issues related to claim (4), the claim about the role of the physician.

There are two issues that I want to address. The first is the role of objective probabilities based upon severity of illness measurement schemes in the physician's information-providing role. The second is whether there are some cases (cases of futility) where the physician's role changes to that of decision maker rather than that of information provider and adviser. Both of these issues have provoked much valuable discussion in recent years.

Many groups have been active in the development of scales for measuring the severity of the patient's illness. These scales have been developed for many reasons, including assessing quality of care (how do the outcomes in different institutions differ when they are treating equally sick patients), providing for proper reimbursement (should institutions receive greater reimbursement for patient A rather than patient B because patient A is sicker), and objectively assessing a patient's prognosis (what is the probability of a patient with a specific severity of illness surviving). Our interest is, of course, in this last purpose, and I will focus on one such scale, the APACHE III Severity of Illness scale (the 3rd. version of the Acute Physiology, Age, Chronic Health Evaluation scale). This scale has been developed over many years by a group at George Washington University Medical Center headed by Dr. William Knaus, and is fully described in a recent paper (Knaus, 1991).

The APACHE III scale has been developed for assessing the severity of illness of adult patients in an ICU. Three factors are relevant to assessing the severity of illness of the patient. Patients receive a specified number of points for various physiological abnormalities (e.g., 8 points for a pulse less than 40 beats per minute), for their age (e.g., 24 points for being 85 or older), and for specific chronic comorbidities (e.g., 11 points for having metastatic cancer). The more points received, the greater the severity of illness. The resulting score, which can range from 0 to 299 points, is the patient's APACHE III score. That score is combined with information about the patient's major disease (e.g., congestive heart failure) and location before admission to the ICU (e.g., a regular hospital room) to yield an estimate of the statistical probability of survival to discharge from the hospital. An example of how this works involves a 56 year old woman (5 points) with acute leukemia (10 points) admitted to the intensive care unit after a nonoperative aspiration pneumonia. The patient had significant physiological abnormalities totalling 92 points. Given a total APACHE III score of 107, given the major disease category of aspiration pneumonia (a relatively unfavorable disease category), and given the patient's location before admission to the ICU in a hospital bed, the APACHE III formula predicts a hospital mortality rate of 81.2 percent.

How good are these estimates? One (traditional) way of thinking about this question is to suppose that patients are predicted to die in all cases in which the APACHE III equation gives them a 90 percent hospital mortality rate; they are predicted to live in all other cases. If that approach had been adopted in the 17,440 patients they studied, this would have led to predictions of 424 deaths with 392 patients actually dying (a predictive value positive of 92.5%) and to predictions of 17016 patients surviving with only 14,384 actually surviving (a predictive value negative of 84.5%). This result compares favorably with the use of other objective schemes and with the subjective impressions of clinicians, and has led to much enthusiasm about APACHE III.

There are many issues that can be raised about this approach; let me summarize some that I raised elsewhere (Brody, 1989): (1) all predictions of mortality using such scoring systems are based on a data base summarizing the previous outcomes of such patients. Clinicians using such predictions must, therefore, exercise great caution in cases where there have been significant medical developments that might change future outcomes and the developers of such measures must be careful to keep their data base upon which the predictions are based up to date; (2) those who have developed and validated these measures have supposed that the advise that will be given is binary advice (the patient will live or die) and that the advice should be conservative (predict death only when the probability of its occurring is very high). There is an alternative approach that I think is more appropriate. What should be shared with the patient is the objective probability of survival. On this approach, the patient decides how conservative they want to be in light of their values; (3) this system, and most like it, provides objective probabilities of one outcome, survival to discharge from the hospital. But patients, in weighing the benefits and burdens of various therapies, need information about other outcomes such as how long they will live after discharge and what will be their quality of life before and after discharge. Until this type of information is provided, clinicians will not be helped in fulfilling their informational role by the development of severity of illness measures and of objective probabilities of survival based upon them.

Despite these issues which still need to be addressed, this approach offers a great deal of promise for more informed patient decision making. Similar approaches are needed for pediatric and neonatal intensive care unit patients. What types of decisions will they lead to? Will we find more patients or their surrogates foregoing ICU care when the probabilities of a successful outcome are low or will we find patients insisting that care be provided because of even a small chance of a successful outcome, a chance whose existence is confirmed by the objective probability data? Studies involving interviews with former ICU patients (Danis,1988) suggest that patients will want intensive care even with only a modest life prolongation resulting from it. A recent study in France (Knaus, 1990) suggests that the provision of objective information leads to decisions for less aggressive therapeutic measures, but the information in that study was given to the clinicians and not necessarily to the patients, so its implications for patient decision making are unclear. This is obviously a matter for further research.

We turn now to the second question identified above, whether there are some cases (cases of futility) where the physician's role changes to that of decision maker. The intuitive idea behind that suggestion is that there are some cases in which it is so clear that the intervention offers no benefits to the patient that it is permissible for the physician to unilaterally withhold and/or withdraw that intervention without consulting the patient (perhaps without even informing the patient of the decision). Without addressing the subtle issues of patient communication in such cases, the Task Force on Ethics of the Society of Critical Care Medicine recently (Task Force, 1990) affirmed such a position when it said that, "A healthcare professional has no obligation to offer, begin, or maintain a treatment which, in his or her best judgment, will be physiologically futile".(p. 1436).

There is little doubt that this appeal to futility arises out of the understandable feeling on the part of many clinicians that patients and their families often demand aggressive therapeutic interventions when they are no longer appropriate(Luce, 1992). But is this just a reflection of a value difference, so that one can argue that physicians must be respectful of the values of the patient whose life is at stake? Or can one define some value-free notion of futility to which physicians can appeal? The concept of 'physiological futility' is an attempt to do precisely that, since it seems far less value-laden than concepts of futility applied to patients (e.g., vegetative patients) whose quality of life is the basis of the judgement of futility.

A good example that has attracted several excellent recent discussions is cardiopulmonary resuscitation and admission to an intensive care unit of patients with metastatic cancer who have suffered a cardiopulmonary arrest. A recent review of the literature (Faber-Langendoen, 1991) reveals no survivals to discharge of such patients. But some of them definitely survive for a period of time in the intensive care unit. Is that transient survival sufficient to make the treatment non-futile or is this still a case of physiological futility? Faber-Langendoen argues that this treatment is still futile because "A treatment that transiently prolongs life in the hospital, particularly the life of a patient dying of an irreversible disease, is without benefit with respect to regaining health, controlling the effects of disease, or relieving suffering"(p.239). Others (Truog, Brett and Frader, 1992, p. 1561) see it differently, arguing that "patients or family members may value the additional hours of life differently". My own judgement is that this second approach is correct, unless one can argue that there is some objective theory of the good according to which the patient or family judgement about the value of the extra hours of life can be dismissed as irrelevant and mistaken (Brody, 1992).

It is important to note that all of the discussions in this section about the appropriateness of the use of therapeutic interventions in the intensive care unit have not referred to the question of the resources required to provide those interventions. Our focus has been on understanding the meaning of benefits and burdens in a world of uncertainty. But economic burdens are real burdens, and I will turn in the next section to a consideration of the additional issues raised by the introduction of considerations of resources.

III. RESOURCE LIMITATIONS AND THE USE OF INTENSIVE CARE

We pointed out above that it is important to distinguish the triaging context, where there are more patients who could benefit from admission to intensive care than there are intensive care beds, from the cost-containment context, where clinicians are being asked to control costs by limiting the use of intensive care even though beds are available. In the first context, the denial of intensive care to some patients who might derive some benefit from it (rationing) is inevitable, and the only question is how such care should be rationed. In the second context, the real issue is whether or not the rationing should take place. We will discuss each issue separately.

Several studies have appeared which analyze how intensive care beds are rationed in the triaging context. While actual practice does not determine what should ideally be done, it is at least helpful to begin by seeing what is actual practice.

The first of the studies (Singer, 1983) was made possible because a nursing-shortage at the Massachusetts General Hospital in the second half of 1981 resulted in a decrease in ICU capacity from 18 beds to 8–14 beds. As a result, there were no ICU beds available for new patients 45 percent of the time, as opposed to only 5 percent of the time until then. The number of ICU admissions declined from 122 per month to 95 per month. Obviously, some patients who would otherwise have been admitted to the intensive care unit were not being admitted. Which patients were not being admitted? The classical concept of triaging, derived from the military setting (Winslow, 1982), involved withholding care from patients at both end of the spectrum, those who were least sick (and would probably survive without the care) and those who were most sick (and would probably die even with the care). Singer found, by way of contrast, a pattern of partial triaging. There was a definite decline in the number of admissions of the least sick patients, those with chest pain admitted to the intensive care unit for monitoring. The percentage of patients admitted to the hospital with chest pain who were admitted to the ICU fell from 85 percent to 63 percent, and those who were admitted were transferred out of the ICU after fewer days. On the other hand, Singer found no decline in the number of the sickest patients admitted. This partial triaging produced no increased mortality in the chest pain patients admitted to the floor rather than to the intensive care, although more patients admitted to the floor needed subsequent emergency admission to the ICU, but even Singer concedes that this does not mean that this partial triaging produced an optimal use of the resources of the ICU. In the end, then, his most important finding was that physicians did not choose a first-come first-served approach; they chose instead a partial triaging approach to the rationing of ICU beds.

A similar pattern was found in a second study (Strauss, 1986) which examined these issues as they arise outside of a crisis situation. The study examined admission and discharge patterns to an 18 bed intensive care unit, and correlated those patterns with the number of beds available at the time of admission and of discharge. Patients admitted when only one bed was available had to be sicker (as measured by their APACHE score or by the likelihood of their having a subsequent myocardial infarction) than patients admitted when two or more beds were available. Similar results were found for the discharge of patients from the unit to the floor. There was no evidence of a denial of admission to the very sickest patients. Again, the authors conceded that they could not endorse this pattern because "it is possible that individuals denied admission when the ICU was crowded suffered increased morbidity and mortality that would not have occurred when more beds were available" (p. 1146).

The most recent study of this phenomenon (Marshall, 1992) did not confirm this pattern. It found, instead, rationing based upon the political power of different surgical services and on various economic considerations. It remains to be settled by further research whether that disturbing pattern is common or is due to special institutional factors.

A number of different attitudes towards these rationing policies have been advocated in the literature. Kalb and Miller (1989), concerned about prognostic uncertainties and about loyalty to patient problems, advocate adopting rationing-avoiding strategies. For surgical patients, this means deferring elective operations until more post-surgical ICU beds become available. For medical patients whose admissions are rarely scheduled, patients should be assigned to step-down telemetry units which are particularly appropriate for chest pain monitoring patients and for those whose major need is intensive nursing. For all patients, the focus should be on earlier discharge, since discharge is often delayed for non-medical reasons. Other authors (Swenson, 1992) have argued instead for the need for principles of justice for the rationing of ICU beds in time of scarcity. Adopting a Rawlsian approach to questions of justice, Swenson argues for the following rationing priorities that do involve full triaging, rather than the partial triaging noted above, although the order still suggests a bias for the most sick even though they are very unlikely to survive: (1)Patients who probably would die without intensive care but who stand a reasonable chance of recovery with that care; (2)Monitoring patients at high risk for life threatening complications who would benefit from the more rapid response to those complications in the ICU; (3)Comatose patients whose expectation of neurological recovery is poor; (4) Patients whose likelihood of survival is poor; (5)Monitoring patients at low risk for life-threatening conditions. A somewhat similar scheme, but without the useful distinction among the monitoring patients and with a reversal of (3) and (4) was advocated by a Task Force of the Society for Critical Care Medicine (Task Force, 1988).

A very special position in this debate is occupied by Engelhardt and Rie (1986), for reasons to be explained below. They begin their analysis by developing what they call an ICU Entitlement Index, a formula designed to capture the elements relevant to deciding about whether a patient should be admitted to an intensive care unit. The formula is PQL/C, where P indicates the probability of a successful outcome, Q indicates the quality of that success, L indicates the length of life remaining (presumably, the length of life in which the success is enjoyed), and C represents the cost required to achieve that success. They use this formula to rank cases from those in which there is an unconditional nonentitlement to ICU care (as a matter of policy, ICU care should not be provided, even if the resources of people and equipment are available), to those in which there is a conditional entitlement to ICU care (care should be provided if beds are available), to those cases in which there is at least initially an entitlement to ICU care, although that entitlement can disappear as the patient's condition worsens, lowering the values of P,Q, and L or raising the value of C.Their argument is that rational people would accept such an approach as a reasonable way of deciding in advance how much society should invest in the provision of health care as a form of insurance against illness.

Several points should be noted about this approach. To begin with, it argues that care should sometimes be withheld even if it is demanded by the patient (or by the surrogate), even if the ICU resources are available, and even if the care is not physiologically futile. This will occur in cases in which there is an unconditional nonentitlement to ICU care because of a low index due to the fact that Q and L are low and C is high and despite the fact that P is high (P's being high makes the treatment not physiologically futile). In this respect, they go beyond all of the advocates of appeals to futility discussed above. Secondly, they offer a justification for such a policy. It is that the very patient now demanding care would have agreed as a rational person to such a policy if asked in advance. Whether or not this justification is successful depends at least in part on whether these postulated agreements in advance (ex ante) to not receive the care should take precedence over the actual demands for the care at the later time (ex post) when the patient is sick. Finally, and most crucially for our purposes, their discussion merges a consideration of the triage context with a discussion of the cost containment context. The very same argument generates the ICU Entitlement Index which provides the basis both of triaging decisions and of cost-containing unconditional nonentitlements.

With that in mind, we turn from the triaging context to the cost-containment context. The introduction of that context leads us to an important traditional argument that would conclude that physicians should not participate in rationing on the grounds that this violates their commitment of loyalty to individual patients. I would claim that if this argument is accepted, it would also lead to the conclusion in the triaging setting that patients who could benefit even marginally from ICU care and who wish that care should be admitted on a first-come first-serve basis until there are no beds left and that the most moral physicians can do is to adopt the strategies described above which were advocated by Kalb and Miller.

This traditional argument has been advocated by major authors (for example, Fried, 1975, Angell 1985). An excellent presentation of it is to be found in the following quotation from Angell:

> As individual physicians, we must do the very best we can for each patient. The patient rightly expects his physician to act single-mindedly in his best interests. If very expensive care is indicated, then the physician should do his utmost to obtain it for the patient (p. 1206).

This type of claim is defended by appeals to the very nature of the patient-physician relation and to the argument that patient trust will be lost unless physicians adhere to this standard.

Several possible responses to this argument have been presented. One, implicit in the Engelhardt-Rie position, is that considering issues of cost is being loyal to the patient's *ex ante* interests in the control of medical costs and is therefore not incompatible with acting single-mindedly in the patient's interests when those interests are understood in an appropriately broad manner. A second (offered in Brody, 1987) is that in a time in which health care, whether provided in the public or private sector, is explicitly or implicitly being paid for by public funds, this single-minded devotion to the individual patient will lead to injustices in the allocation of these public funds.

It seems to me, however, that there is another question to be considered. Is the physician morally free to take this single-minded approach to promoting the interests of the individual patient? Is any one of us ever morally free to attend to only a single set of considerations? I believe that the answer to these questions is that the physician, like the rest of us, is not morally free to adopt this approach. We all live in a web of loyalties, all of which make demands upon us, demands which can be conflicting. We have loyalties to ourselves, to our family and friends, to our broader community, to our religion, to those with whom we have entered into a variety of relations, *etc.* Each of these loyalties generates moral claims upon us, claims that are often conflicting. Physicians, like the rest of us, must learn how to balance these claims rather than to pay single-minded attention to any one of them. The physician who may, *quite legitimately*, say that I cannot see for a time any more indigent patients (having already cared for them in the past) because my loyalties to myself and my family call upon me to see at this point more revenue-generating patients can hardly then say that he or she cannot participate in helping society contain costs because of a single-minded loyalty to patients.

IV. CONCLUSIONS

This paper has attempted to identify and clarify a variety of issues about ICU care, rather than to argue for definitive conclusions about them, although I have certainly felt free to indicate my own leanings on a variety of these issues. So a traditional summary of conclusions is not in place. But I do want to emphasize two points. The first relates to a theme emphasized at the end of section I. I noted there that given the wide variety of patients in intensive care units, and given the wide variety of interventions provided in such units, it would not be helpful to discuss ICU interventions in general without specifying which interventions for which patients. That theme, I believe, has been reinforced by what we have seen in the other sections of this paper. The use of APACHE III for prognostication and the development of triaging criteria, to mention only two examples, requires great attention to the clinical details of different cases. In this way, an ICU casuistry may emerge. The second is a reminder about the benefits of intensive care. In many discussions, I believe, too much attention is devoted to fighting the technological imperative and to emphasizing only the legitimacy of limiting care. But as the data from the neonatal studies and from APACHE III show us, there are also cases where intensive care is of great benefit and should be encouraged. We need to make sure that our casuistry of ICU care is a balanced casuistry, one which recognizes its many benefits as well as its many costs.

REFERENCES

Angell, M.: 1985, 'Cost Containment and the Physician', *JAMA*, 254, 1203–1207.

Brody, B.A.: 1987, 'Costs and Clinicians as Agents of Patients', in Brody, B.A. and Engelhardt, H.T. (eds.) *Bioethics*, Prentice Hall, Englewood Cliffs, N.J.

Brody, B.A.: 1989, 'The Ethics of Using ICU Scoring Systems in Individual Patient Management', *Problems in Critical Care*, 3, 662–670.

Brody, B.A.: 1992, 'Special Ethical Issues in the Management of PVS Patients', *Law Medicine and Health Care*, 20, 104–115.

Bryan-Brown, C.W.: 1992, 'Pathway to the Present: A Personal View of Critical Care' in J. Civetta, R.W. Taylor, and R.R. Kirby (eds.) *Critical Care*, 2nd. ed., J.P. Lippincott Company, Philadelphia, 5–11.

Danis, M. *et al.* : 1988, 'Patients' and Families' Preferences for Medical Intensive Care', *JAMA*, 260, 797–802.

Engelhardt, H.T. and Rie, M.A.: 1986, 'Intensive Care Units, Scarce Resources, and Conflicting Principles of Justice', *JAMA*, 255, 1159–1164.

Faber-Langendoen, K.: 1991, 'Resuscitation of Patients with Metastatic Cancer: Is Transient Benefit Still Futile', *Archives of Internal Medicine*, 151, 235–239.

Fried, C.: 1975, 'Rights and Health Care', *New England Journal of Medicine*, 293, 241–245.

Groeger, J.S. *et al.*: 1992, 'Descriptive Analysis of Critical Care Units in the United States', *Critical Care Medicine*, 20, 846–863.

Hack, M. *et al.*: 1991, 'Very Low Birth Weight Outcomes of the National Institute of Child Health And Human Development Neonatal Network', *Pediatrics*, 87, 587–597.

Hastings Center: 1987, *Guidelines on the Termination of Life-Sustaining Treatment and the Care of the Dying*, Hastings Center, Hastings-on-Hudson, N.Y.

Knaus, W. *et al.*: 1990, 'Do Objective Estimates of Chances for Survival Influence Decisions to Withhold or Withdraw Treatment', *Medical Decision Making*, 10, 163–171.

Knaus, W. *et al.*: 1991, 'The APACHE III Prognostic System', *Chest*, 100, 1619–1636.

Luce, J.M.: 1992, 'Conflicts over Ethical Principles in the Intensive Care Unit', *Critical Care Medicine*, 20, 313–315.

Marshall, M.F.*et al.*: 1992, 'Influence of Political Power, Medical Provincialism, and Economic Incentives on the Rationing of Surgical Intensive Care Unit Beds', *Critical Care Medicine*, 387–394.

O.T.A.: 1981, *The Costs and Effectiveness of Neonatal Intensive Care*, Office of Technology Assessment, Washington, D.C.

O.T.A.: 1984, *Intensive Care Units: Clinical Outcomes, Costs, and Decisionmaking* Office of Technology Assessment, Washington, D.C.

O.T.A.: 1987, *Neonatal Intensive Care for Low Birthweight Infants: Costs and Effectiveness*, Office of Technology Assessment, Washington, D.C.

President's Commission for the Study of Ethical Problems in Medicine: 1983, *Deciding to Forego Life-Sustaining Treatment*, Government Printing Office, Washington, D.C.

Singer, D. *et al.*: 1983, 'Rationing Intensive Care-Physician Responses to a Resource Shortage', *New England Journal of Medicine*, 309, 1155–1160.

Strauss, M.J. *et al.*: 1986, 'Rationing of Intensive Care Unit Services: An Everyday Occurrence', *JAMA*, 255, 1143–1146.

Swenson, M.D.: 1992, 'Scarcity in the Intensive Care Unit:Principles of Justice for Rationing ICU Beds', *American Journal of Medicine*, 92, 551–555.

Task Force on Guidelines, Society of Critical Care Medicine: 1988, 'Recommendations for Intensive Care Unit Admission and Discharge Criteria', *Critical Care Medicine*, 16, 807–808.

Task Force on Ethics of the Society of Critical Care Medicine: 1990, 'Consensus Report on the Ethics of Foregoing Life-Sustaining Treatments in the Critically Ill', *Critical Care Medicine*, 18, 1435–1439.

Truog, R.D., Brett, A.S., Frader, J.: 1992, 'The Problem with Futility', *New England Journal of Medicine*, 326, 1560–1564.

Winslow, G.R.: 1982, *Triage and Justice*, University of California Press, Berkeley.

MICHAEL A. RIE

ETHICAL ISSUES IN INTENSIVE CARE: CRITERIA FOR TREATMENT WITHIN THE CREATION OF A HEALTH INSURANCE MORALITY

INTRODUCTION

Critical care medicine involves the application of medical strategies, drugs, and technologies to support the vital organ functions of humans in high technology environments that customarily form a significant portion of the modern hospital's budget and manpower (Russell; Berenson; Rie; Engelhardt). Prior to 1960, there were no intensive care units (ICUs) in occidental hospitals and the resource allocation was non-existent. Patients who may linger for long periods today died swiftly in the earlier period. Understanding the history of the development of critical care medicine is central to understanding the insufficient moral attention that was paid to these issues in the past, and the demand they place upon us now to reconceptualize medical care in the latter 20th Century.

In the United States the legal and public model for withdrawal of life support follows the principle of autonomy. Most recently the United States Supreme Court, in the matter of Nancy Cruzan (Cruzan), decreed that individual States had the authority under the Constitution to set the standards of evidence for determining patient's wishes. The failure to leave instruction that meet the standards for the state would result in a duty of physicians to continue support for those in chronic vegetative coma. Since 1976 (Quinlan) critical care units have found themselves the object of public uncertainty with unbridled interference by moral interlopers in the private fiduciary trust relationship that characterized the sensitive and humanistic relationship between a physician and the patient since the days of Hippocrates (Rie, 1992). Not surprisingly, these issues of individual autonomy are central to questions about how high technology medical care should be developed particularly in cases where individuals are unlikely to benefit from that care.

K. Wm. Wildes (ed.), Critical Choices and Critical Care, 23–56.

The second theme for our discussions will be to what extent the failure of modern religious institutions to forthrightly address the moral dimension of health resource allocation in secular health policy constitutes a genocidal social threat to the existence of the greater society and a failure to contain health care as a relative human need. Religions are able to provide moral value content in a universe of secular pluralism.

Health care is a rapidly increasing cost in all western countries (Schieber and Poullier). There is an ever increasing percentage of societal wealth that goes to the provision of medical services. In the United States health care has increased progressively since the 1960s from levels of about 6% of the gross domestic product to present levels approaching 14%. At least 1% of the Gross Domestic Product (GDP) of the United States is expended in ICUs (Berenson) and the present figure may approach 1.6% of GDP. With an increasing geriatric population there is increased demand for services because of increased utilization of services in acute hospital care by older citizens. Heart surgery was unknown for individuals over the age of 75 or 80 in the 1950s. Today repeat coronary artery surgery and valvular heart surgery for individuals over the age of 80 or 90 is not an uncommon event. The increased exposure of octogenarians to high technology medicine may explain the slightly increased life span of Americans at age 80 relative to German, English or French citizens. Further, the mass media have educated the public to believe that such procedures are possible and are the fruits and miracles of modern science. It is not only avaricious physicians preying on patients to have surgery and to make money. There is simultaneously an unquenchable demand from consumers and a pandemic cultural desire to fight mortality. Health care is such an engaging topic because people want it for themselves and are unwilling to face the reality of personal finitude and the tremendous costs that are associated with delaying mortality near the end of life.

I. THE BEGINNINGS OF CRITICAL CARE: THE MEDICAL CULTURAL MANDATE TO TREAT PHYSIOLOGY APART FROM PERSONS

As with most technologic medical endeavors, critical care medicine found its origins in the needs of desperately ill people who were able to benefit from something new. In the 1930s the endotracheal tube was developed. It is a tube that can be passed through the larynx and into the windpipe. This tube allowed for air or oxygen to be injected under pressure into the lungs and for the lungs to remain inflated while a surgeon was able to open the chest and operate upon the lungs. Prior to that time operating on the lungs was associated with a high risk of death from asphyxia. From the 1930s to the 1950s the use of an endotracheal tube was principally restricted to the surgical arena and it was not used after surgery. In 1950–51 an epidemic of polio myelitis spread through northern Europe and many children were dying of respiratory failure due to paralysis of the muscles affecting the chest. As chronicled by Pontoppidan and co-workers (Pontoppidan), a Danish anesthesiologist, Ibsen, was first to show that by placing an endotracheal tube into the windpipe of a child with polio myelitis that the patient, who was otherwise dying, would suddenly have revival, that the mind would be preserved, and the rest of the body as well. As polio myelitis involving the respiratory muscles usually is reversible with time from natural biologic healing, children who otherwise would have died were saved. There ensued a period of inquiry and technologic advance in which "human ventilators" in the forms of medical students and others would go to the bedside of children with polio and squeeze bags of oxygen. Shortly thereafter Engstrom in Sweden developed the first mechanical ventilator.

Soon after the demonstrated success of the endotracheal tube, the human ventilator, and the mechanical ventilator to the resuscitation of children with polio, it became clear that there were many technical complications to overcome. The 1950s and 1960s saw the remarkable development of the mechanical ventilator, better endotracheal tubes, and tracheostomy tubes that would not injure the windpipe and humidification systems to see that the lungs would remain with an adequate quantity of water such that the mucous produced by the bronchial linings would not impact and clog the windpipes. To affect the application of this technology to the patient required that teams of people be assembled as the amount of labor involved in sustaining this process was significant. As well, the monitoring systems available to assure that oxygen would enter the lungs and carbon dioxide would exit to sustain the process of cellular respiration were very primitive. Many accidents occurred in the early years requiring a high intensity labor component. Lastly, the pressure effects of passing gas into the lungs by the ventilator results occasionally in disruptions of the anatomy of the lungs and these also required further technical advances.

Not surprisingly, the capacity to resuscitate human life was an intoxication of mammoth proportions to the medical and nursing professions. By 1960 academic medical centers in Toronto, Baltimore, Scandinavia and Boston had developed to further study and expand the usage of these technologies. Shortly after this period it was observed that patients undergoing open heart surgery would improve in their post-operative survival by using the mechanical ventilator to diminish the work of the muscles necessary for breathing while the heart was allowed to repair itself after the operation.

The arrival of the mechanical ventilator then led to an expanded definition of what kind of patients could conceivably survive a surgical operation. Elderly patients who previously would have automatically died due to insufficient blood flow to the intestines associated with vascular disease and in the presence of severe heart and lung disease were slowly but irrevocably being considered as candidates for operations in the 1960s.

The 1970s saw the explosive period in which these populations, and even younger extremely sick people who otherwise would have died for lack of surgical care, were being offered surgery and in many instances surviving with restoration of high quality functional life. Unfortunately, as these successes developed, there was the parallel creation of a group of individuals within these populations who would theoretically have the capacity to benefit from the biological success of removing a diseased or obstructed intestine but whose underlying medical circumstances and chronicity of disease would lead to post surgical outcomes that would require the prolonged care of the patient in the intensive care unit. Indeed, academic ICUs refined the process of critical care in ICUs to the point where even many of these patients who otherwise were dying in less sophisticated centers were regularly living and often achieving high qualities of life free of technology after more prolonged stays in the ICU. As young physicians were observing and training in these centers, practice patterns were transferred to community hospitals. The 1960s and 1970s saw the rise of cardiopulmonary resuscitation (CPR) and the practice of administering CPR in all cases unless patients had previously declined this option. Despite the universal poor results of CPR (Murphy) in preterminal or far advanced chronic disease (Younger), physicians remain fearful of being accused of malpractice for failing to attempt resuscitation for individuals who had not given prior authorization to forego this therapy.

We have begun with a detailed analysis of the early development of critical care medicine to convey the tremendous emotional context within which young physicians and nurses are trained universally. The capacity to resuscitate somebody from cardiac arrest is as compelling for the physicians in the world over. The physician is as much wrapped up in the project and morality of fighting death as is the patient's desire. However, the recent rise of cost containment as a driving force in health care allocation and the need to contain health care expenditures as a risk to other societal endeavors forces us to examine the mores and the conjoint activity of patients and physicians. If the physician is to dedicate his individual talents and energies to the care of the patient and to hold the patient interests above none other, then it is clear that asking the individual physician in the care of patient to be responsible for the allocation of resources constitutes a conflict of interest at the highest level of morality that must be addressed publicly. We then fall into the usual predicament of how would we arrive at a publicly articulated value that would be acceptable to members of secular pluralists societies? Unfortunately, our economic and health policy literature is full of examples of how to avoid this clear and uncomfortable statement. For example, Ashby and Craig in addressing the question of "why do hospital's cost continue to increase"? arrive at the conclusion: "Our findings suggest that cost containment efforts should be concentrated on controlling increases in intensity of services . . . Gaining further ground in controlling service intensity will require extensive involvement by hosptials and payors in monitoring and influencing physicians practice patterns". A reflective moment on this type of statement (which is almost universal in the health economics literature) leads to the moral reflection that the physician is a vulnerable moral agent whose thoughts and behavior are now the subject of unarticulated public policy. While the Hippocratic tradition served medicine as a profession for over 2000 years, it is now being used to disadvantage physicians and patients and to focus the failures of absent moral values in our resource allocations as being the result of physician behavior and selfish and greedy human desires to fight mortality by patients.

II. INTERNATIONAL DIVERSITY IN CRITICAL CARE MEDICINE: RESOURCES, GOVERNMENTS, PRACTICE, AND HYPOCRISIES

For many years the OECD has published data on international comparisons of health resource allocations in varying countries within the OECD (Schieber and Poullier). It is apparent that the data assembled by OECD is being attacked by others because standardized methods are not used.[1] Within the great diversity of money allocated to health care across nations, the role of ICU expenditures is buried deeply in the data and the specifics are largely unknown in the public literature. What is known is that the use of intensive care varies widely across the Western nations while the end results of longevity and health care seem to be remarkably unrelated to national expenditures of money. Thus, the United States has a high allocation in hospitals for the use of intensive care beds and the national average approximates 6–7% of general hospital beds. By comparison, intensive care units are not as prevalent in Great Britain and other European countries where the average use of intensive care beds approximates 2–4% of total hospital beds. It is estimated that roughly 20% of a hospital's budget in the United States goes to the care of critically ill people. As there are no moral strictures on the entitlement of patients to consume ICU resources and cost containment has been applied to other discretionary areas of the hospital budget, it is unquantitated at the present time in the latter 1980s and early 1990s how much critical care is increasing as a proportion of the general hospital budgets. Largely, this data is hidden and in the United States it is truly unavailable because hospital financial officers will not share publicly the true nature of cost as this is privileged public information that the government and payers would like to acquire.

On the other hand, we are able to see that there are now objective methods for assessing prognosis of patients who are likely to survive or die from critical illness. In 1981, Knaus and co-workers were the first to describe a scoring index known as APACHE I which stands as an acronym for Acute Physiology and Chronic Health Evaluation index which is a physiologically based classification system. Throughout the 1980s this system has now been refined into a third generation form called APACHE III (Knaus et al.). In addition there has been a European modified version developed by LeGall and co-workers entitled SAPS I and SAPS II, an acronym for Simplified Acute Physiology Score (LeGall et al; 1984; LeGall et al., 1993). There is also the MPM II international ICU data base created from seventeen European and North American countries (Lemeshow et at.). These systems have now been employed with large scale computerized data acquisition such that it is possible, at least statistically, to objectively characterize the prognosis for death in the short term for patients who have severe physiologic problems even when care is applied with full resource allocation and the best cognitive performance that can be achieved by human physicians. What emerges is a picture of an aging and increasingly sick population of desperately ill humans which consume resources in bimodal distribution. One mode is characterized by short term use of intensive care with physiologic support and monitoring where biologic healing of the host patient is highly probable. These resource allocations tend to be much cheaper per capita than for the second mode. The second mode is characterized by individuals with multiple organ dysfunction and chronic infirmities often associated with advanced age in which there has been an elective decision to perform heroic life-saving procedures which circumscribe and develop the "definition of the need" for intensive care. Once the patient enters the intensive care cycle there is a medical culture rather than a proven outcome assessment of technology usage that compels physicians to apply invasive technology which may then lead to secondary iatrogenic complications. The result in the second mode is a group of individuals who consume many resources over several weeks or months who have a statistically poor prognosis that may culminate in death or serious disability.

More importantly, the use of the objective prognostic determination scores allows for international comparison of what different countries attach to the monetary value of the human life of a citizen as compared to American ICUs. To graphically show this, let us examine Figure 1 which is a mortality rate adjusted severity of illness scale comparing the Cedar-Sinai Medical Center in Los Angeles to a combined study of French Intensive Care Units for elective surgery (Shabot, et al.). What can be seen from this previously unpublished comparison of intensive care results in Los Angeles versus France is that there is a minor difference in outcomes when objective comparisons are applied to low intensity patients of SAPS scores of 10 points or less. As one moves into higher intensity levels of illness, one observes a significant difference in death that occurs in French ICUs as compared to Los Angeles although the rate of death rises in both groups. Herein we can see the subtle public policy issue manifested in two different systems. In France, public expenditure

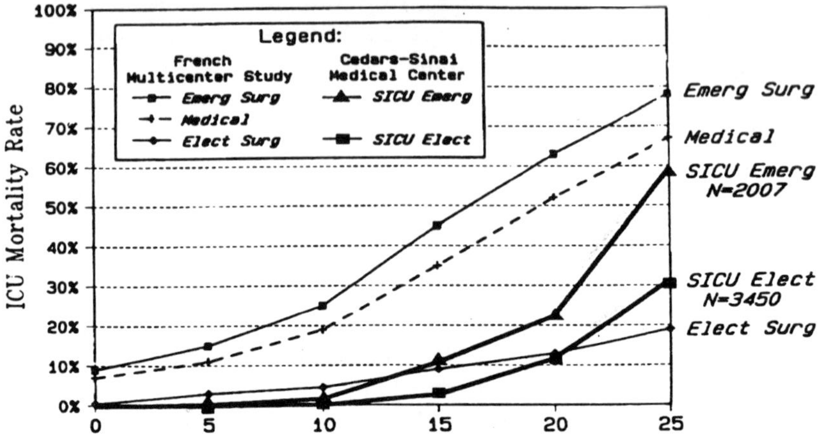

Fig. 1. Severity adjusted survival comparison of American and French ICU outcomes for abdominal surgery. The x-axis represents SAPS-I scores. The differences are statistically significant at P = 0.02 for scores of 10 points and P = 0.0005 for scores of 15 and 20 points [20].

limits the capacity to acquire technology and labor. Therefore ICUs are under greater pressure to make choices and to function in the triage mode in which physicians are forced to select which patients will and will not be candidates for intensive care or when the intensity of effort declines. One sees an implicit rationing system in which the dying occurs faster however. There is no public articualation of the policy because of our failure to understand the limits of mortality and of technology application. On the other hand, in Los Angeles we see lower levels of mortality in which some individuals are surviving who otherwise would have died in the French intensive care unit but where the allocation is much larger and the sanctity of life principle is more operant in the way in which the system chooses to allocate its financial resources. The fact that individuals survive to leave an ICU does not mean that they will be alive to benefit from the fruits of a quality of life that would have made such survival worthwhile. This was graphically demonstrated as early as 1974 by Cullen and co-workers (Cullen, et al.; Keene and Cullen) who developed the Therapeutic Intervention Scoring System (TISS) which was an index quantitative comparison of patient care. Cullen and Civetta were perhaps the first to point out the inverse relationship between resource consumption and survival in intensive care units.

Though this is a well understood phenomenon (and we now have much better quantitative data through the APACHE III system and others to document that the patients who consume the most resources are likely to have the most marginal quality of survival at the highest price), there is no evidence that physicians are bending to this data in terms of the care of individual patients.

III. RESOURCE NON-ALLOCATION TO SPECIFIC CONDITIONS: THE NAZI PRESUMPTION

Our discussion would be incomplete without an accounting of how nations make choices about the types and quantity of intensive care they wish to purchase. There is generally a commitment to provide adult intensive care unit services both for medical and surgical problems throughout Western Europe and the United States. However, the use of intensive care for newborn babies and those who have not yet been born suggest that there are broad national moral choices and economic choices that have been made which represent major differences in how western nations choose to allocate their resources. In the United States there is a large quantity of intensive care resources that goes for children of extreme prematurity the results of which may require chronic intensive care at the beginning of life. Saving children who have been injured severely in their brains from premature development and hemorrhage creates a population of children who require sustained care on the ventilator often in quasi intensive care environment at home.

With the rise of AIDS and with the large number of premature children being born and abandoned by cocaine addicted AIDS mothers, stress is placed on public budgets to care for children and stress the availability of services for those who have a higher chance of survival. There is an unending debate on this subject in the United States which was most clearly captured by the Oregon legislature's decision not to fund care for children born below 500 grams of weight. In America this is morally debated in politics by asserting that the denial of such services constitutes discrimination against those with disabilities. President George Bush of the United States declined to authorize governmental funds to be employed in Oregon on the basis that Oregon's decisions in health care, of which neonatology was a prime example, constituted discrimination against the disabled as was developed in national rights based legislation called the Americans With Disabilities Act.

On the other hand, European nations and Israel take a very different view of these matters. The same applies to intensive care resource allocations for AIDS (Rie 1992). As can be seen from these dialogues in *Critical Care Medicine*, raising the issue of developing a forthright morality of disentitlement for those who are not able to benefit from critical care resource allocation so as to enjoy cognitive life or a life of existence beyond a small period of time purchased at great cost, raises what shall be termed "the Nazi Presumption". Repeatedly, we see that attempts to create affirmative disentitlement as a matter of social consciousness and public policy either for AIDS, neonatal intensive care, or adult intensive care at the margin, brings assertions of dictatorships and the extermination of the helpless. The challenge for Catholic theologians is to find a way to rearticulate the social worth of medical care in a balance with individual entitlements and disentitlements as virtuous without alluding to the rise of dictatorships and evil in human history. There are clear distinctions between Nazi Germany and the Oregon health proposal. Our task should be to capture the moral and theologic essence of how we delineate the difference between Hitler and Kitzhaber. Should the delineation be clear, then the moral design of how one vests allocatory power and the limits of technology application follow as a necessity for medical caregivers and patients.

IV. OPEN AND CLOSED ICUS: THE MORALITY OF COMMAND AND CONTROL

In the United States the "Open ICU" predominates as an organizational and economic matter. The "Open ICU" is defined as one in which a physician or surgeon is the fiduciary of a patient and the "public thief" for that patient in terms of marshalling the resources of the insurance systems, governmental allocations, and hospital resources for the care of the patient. The physician exercises control over the care of his patient in an ICU and simultaneously allocates resources. Intensive care physicians exist in this culture and provide consultation at the bedside but authority for who controls the resources as a legal matter remains within the hands of the attending physician.

A second model may be called the "Closed ICU" which exists to some degree in the United States but is the predominant model for European ICUs. In the closed system, a patient who would ordinarily be under the care of the physician in the hospital or community who becomes critically ill and requires intensive care will move to an ICU where the patient will be cared for by a group of physicians and nurses called the "Intensive Care Team". As long as the patient resides in the ICU, the patient will remain under the care of the "team". When the patient is discharged, fiduciary responsibility and Hippocratic duties for the care of the patient will be transferred to other physicians.

In either of these systems there has been a naive bioethical presumption that patients and their families are truthfully appraised of the medical prognosis so that autonomous choices regarding limitations of care could be developed. Surbone (1992) and Pellegrino (1992) graphically depict that the bioethical code for Italian physicians in the Italian culture is not to tell the truth to patients about the nature of terminal illness or untreatable end-stage cancer. Surbone makes a graphic example of how the bioethics taught to Italian physicians is different from that which exists in America with regard to how cancer physicians are educated and taught to practice. At the very least this should throw into transcultural perspective the paternalistic versus the autonomous nature of decision making in health care and the role to which medical or governmental paternalism may govern the distribution of resources. Whether one is dealing with the open or the closed system in this regard, it is apparent that the conveyance of information, the conveyance of prognosis, and the conveyance of medical opinion are major determinants about how individuals will view the proposed further use of resources for either themselves or their family members.

It is significant that despite moral or emotional views attached to the passing of life, resources consumed by the patient are not those of the patient but those of the social organization or insurance scheme that made those resources available. Both of these systems have moral problems when we attempt to balance the needs of society versus the needs of individuals. In the European "Closed ICU", we lack a system of checks and balances and resource allocation and patient care are integrated via global budget restrictions. In the American "Open ICU", we do almost as much except for the tension which exists between intensive care physicians and primary care physicians. Regardless of the administrative structure, the need for the preservation of patient trust in the physician who cares for that patient remains a primary goal of humanistic health care. The need for specific "gatekeeping" of the resources is one that should be clearly separated from that of the fiduciary role.

V. ICU GATEKEEPING: A MORAL POLICY OF AFFIRMATIVE DISENTITLEMENT

There are both moral and non-moral reasons to seek a strict differentiation of professional roles between those who triage or micro-allocate resources to individuals and those who are the physician fiduciaries of patients. If one earns one's livelihood at the bedside of a critically ill patient, one has an interest in seeing that the number of patients and the number of critical care events that are dispensed to patients remain high and do not decrease. If one is dedicated to the converse, one could jeopardize the definition of one's role and one's job. The same would apply to a definition of how many ICUs should exist within a given community to meet the needs of the critically ill in that community (Rie 1989). In general, the United States is characterized by an overabundance of ICUs and an overabundance of critical care physicians. In Europe there is overabundance of physicians to the point of significant physician unemployment in common market nations. In either of these systems, the physician functions more as a marketing agent for patients and their desires to combat mortality than they do as implementors of specific policy. If the Intensive Care Unit Entitlement Index (Engelhardt and Rie) is to have any meaning in terms of developing macro economic health policy for ICUs and simultaneously allocating resources to individuals, then we must address the creation of a horizontally integrated system which would provide the moral elements of a democratically developed ICU allocation. This allocation would be based on an entitlement index where the ability of patients to benefit from the resource allocation was the pre-eminent criterion for allocation. This would apply only to the concept of minimum basic medical resources as it would be presumed that those with more resources would be able to buy a higher level of care. In this articulation we accept the vision that one can only allocate those resources that exist within the insurance system while leaving individuals with private resources free to purchase additional services.

VI. THE INTENSIVE CARE UNIT ENTITLEMENT INDEX: CREATING HEALTH INSURANCE MORALITY IN AN OREGONIAN GRASS ROOTS PLURALIST DEMOCRACY

For definitional purposes, this essay takes Oregonian health policy to encompass those pure elements listed in Table I. These were the original elements of the Oregon Basic Health Serices Act passed into law in 1989.

On February 1, 1994 the Medicaid resource allocation program was implemented in Oregon as well as the less publicly known portion of the legislation, Senate Bill 935 (Chapter 381, Oregon Acts of 1989). Oregon passed legislation which follows the concept of an employer mandate to provide health insurance for large and small businesses. Although the effective date of this mandate has been pushed back into the late 1990's, it is now legal under the laws of Oregon for managed care plans in the private sector to undertake contractual relations with providers and subscribers that allow for explicit limitation of individual patient care within the benefit package. As long as the benefit package at least equals that in the public Medicaid sector, the legal protections that apply to providers in the Medicaid program of Oregon will also obtain for them in the private health insurance sector. This background in the philosophy of law and health insurance relies heavily on the theory of contracts replacing the theory of unstated human rights in health insurance (Havighurst a, 1986; Havighurst b, 1986). The nature of the economic relationships in the law are eloquently detailed by Morreim (Morreim, a 1991; Morreim b, 1991; Morreim 1989).

We previously developed this framework in 1986 (Engelhardt and Rie). In that publication, we laid out the general moral theory for hypothetical contractors possessed of rationality who desired to create a definition of intensive care unit entitlement to benefits based on an Intensive Care Unit Entitlement Index. That index states:

$$\text{“ICU} - \text{EI} = \text{PQL/C”}$$

In this index, P indicates the probability of a successful outcome for a specific condition. Q defines the quality of the successful outcome. L defines the length of life remaining, C defines the costs required to achieve therapeutic success. In that publication, we highlighted several clinical examples. We described a range from absolute entitlement within an insurance system through entitlement leading to subsequent non-entitlement because of a decline in prognosis and potential clinical

TABLE I

FUNDAMENTALS OF A PURE OREGONIAN HEALTH CARE POLICY

- When resources are insufficient, allocate resources not patients.
- In secular pluralist democracies moral authority is acquired in limited measures by negotiation between providers and consumers to prospectively prioritize the creation of a list of health insurance benefits. These policies define the macroeconomic moral response to the natural and social lotteries of life (Engelhardt and Rie).
- Health policy that lacks moral prioritization of beneficial care when resources are insufficient mandates declining standards of care with measurable increases in the morbidity and mortality of patient populations.
- In limited democracies the creation of universal access to health care in a population of unequal personal wealth requires transfer payments to the poor to be limited by the extent to which the majority of citizens consent to the taxation of their personally created wealth.
- People and businesses create wealth. Governments requisition wealth.
- The historical liberal maxim that moral conflict requires moral neutrality while resources are consumed by fiat is abandoned.
- Respect for religious and moral diversity of patients requires public toleration of these expressions of health care desires and their financial support and accountability from private sources of wealth.
- Futile medical care is not a disability under the Americans with Disabilities Act. Futility is a publicly derived, morally contentful definition of unfortunate human circumstances defining the finitude of human life and the limits of our communal resources.
- Resource allocation in critical care medicine fuses responsibility and authority in a contractually limited, medically directed managerial structure. Fiduciary trust relationships between doctor and patient are preserved but limited in the extent of resources those relationships may requisition from the limited resources of a basic health insurance plan.
- All politics are local and health insurance morality that fuses responsibility and authority to ration is regionally created. Federalism and the constitution of 1789 are alive and well in American politics and law.
- Oregon morally prioritizes liberty, equality, access and prosperity in health care. The American constitution of 1789 was silent on this subject.
- Resource allocation policy limits individual entitlement claims in the tort system of law by legislation.
- Unequal health care is officially acknowledged and the basic package of care is guaranteed to all while the freedom to own and use private wealth to buy more health care is preserved in Oregonian health systems.
- The best of care for all is not attainable. The moral dimensions of medical prognostic probability and clinical impression must be monetarized in individual entitlement allocations at the mesoallocatory policy level.

outcome to absolute non-entitlement because of inability to benefit from the resource allocation. In 1986, the science of predicting ICU outcome was in its infancy and the computerized software to carry out rapid data base acquisition was yet to be developed. In the intervening years the predictive models have improved sufficiently so that it may now be reasonable to perform mortality probability estimates for individual patients (with significant caveats).

It is assumed that any group of individuals wishing to voluntarily enter into the creation of a morally contentful health insurance contract would have full access to the science of prognostic scoring and be aware of its limitations. As well, it is assumed that the contract negotiators would have full access to the nature of clinical imprecision and clinical fallibility when physicians are left to their own devices in the practice of medicine. They would have full knowledge of the variability of practice styles, costs, and outcomes.

Since 1986, the Society of Critical Care Medicine (SCCM) has made fundamental changes in the definition of the ethical integrity of critical care medicine as a medical discipline. In 1994, SCCM published its *Consensus Statement on the Triage of Critically Ill Patients* (See Appendix). The consensus statement is precedent setting because it is the first in occidental medicine to specifically call for the limitation of autonomy entitlements and rights of consumption of patients when resources are scarce in a civilian health care setting. The statement strongly supports prognostic scoring instruments as a method for making triage prioritization policy. The consensus statement is offered as an informal guideline to critical care practitioners and suggests (but does not justify) that the locus of policy making resides within a hospital. Finally, the consensus statement lists a series of categories that do not merit ICU admission. These include: (A) Patients who competently decline intensive care or request that invasive therapy be withheld. (B) Patients declared brain-dead who are not organ donors and, (C) patients in a persistent vegetative or permanently unconscious state. The statement further specifies that moral pluralism in the society may result in situations where individuals holding religious or moral convictions may legitimately have a basis favoring the provision of such treatments of the conditions but that the costs of such care should not be borne by the general society and the provision of such services should not foreclose the treatment of other patients who would benefit from critical care. The latter is likely to be of great public policy importance because it acknowledges moral diversity, permits the private ownership of money, and respects the free exercise of religion in health care but does not burden the majority and their financial obligations to a system of health care that is only desired by the minority. For the first time a medical organization supports financial accountability for moral diversity and provides the last element so necessary for the creation of an integrated and monetarized critical care resource allocation system.

A) Health Policy Framework for the 1990's

Shortell and Reinhardt (1992) suggest that the evolution of health policy in the 1990's should consider 7 elements. These are: 1) There is an interdependency that relates to the creation of incremental policy. 2) Health policy should be patient centered. 3) Health policy should articulate the balance between communitarian values in health insurance vs. individual wishes and expressions of autonomy. 4) Health policy should promote explicit clinical management systems. 5) Health policy should promote fiscal and clinical accountability. 6) The policy should have flexibility to encourage innovation in accountability criteria and provider incentives. 7) The policy should anticipate implementation issues, mid-course corrections, and input for those directly affected.

What exactly do we know about explicit resource allocation in critical care? The answer to this question is absolutely nothing. We have a *status quo ante* of health financing, conflicting case law, uncertain rights to nebulous insurance benefits in morally empty insurance policies, and the existence of an advancing science of prognosis research that is not being applied to individual clinical entitlement limitations. In addition, we have a series of sophisticated reflections on ethics in critical care medicine and we have evolved a comprehensive system of ethics committees that act as public sounding boards and dispute resolution loci for resolving disputes in American hospitals. However, when patients object to the removal of therapy on ground of futility or non-medical grounds, they are still able to bring litigation against hospitals and physicians. Accordingly, ethics committees can be obstructed from carrying out any authoritative role with regard to resource allocation.

B) Health Insurance Contracts as the Locus of Moral Authority

The basic construction of the horizontally integrated Oregonian accountable ICU is illustrated in Figure 2. In this figure, we begin with the

Administrative Scheme For Provider Reimbursement Eligibility

Fig. 2. Schematic principle for horizontally integrated critical care resource allocation policy based on the morality of reimbursement criteria. Moral authority is created in the health plan. Institutions and individual providers elect to participate in the morality as expressed in "bylaw symmetry" with the health plan moral limits on individual care and the gatekeeping principle.

premise that moral authority in health insurance is acquired at the time the insurance policy is created. The insurance policy is created and continuously recreated by an on-going conversation among subscribers, providers and payors. However, unlike conventional insurance, the insurance laws of the jurisdiction are rewritten to require that subscriber representatives and provider representatives be part of the creation of the health insurance policy. In this system, the benefit manager of a corporation or business is prohibited from having exclusive control of information to employees. Employees in such a system are not eligible for health insurance until such time as they have had full access and understanding of the benefit package at the time they enter employment. Responsible policy requires individual responsibility. Governmental paternalism in policy creation has failed to address rational decisions. The process must return to the people. On the other side of the equation, the same issues apply to hospitals and to physicians who would admit patients to an ICU.

In this system, neither hospitals nor physicians may qualify as recipients of health insurance payments unless they subscribe to the contractual obligations within the insurance policy. For primary physicians, there will be no possibility to be both a primary physician and also an allocator of resources or critical care consultant. The term primary physician in this chapter does not mean the fields of general medicine such as family medicine, pediatrics, internal medicine and obstetrics and gynecology, but refers to those physicians including surgeons, cardiol-

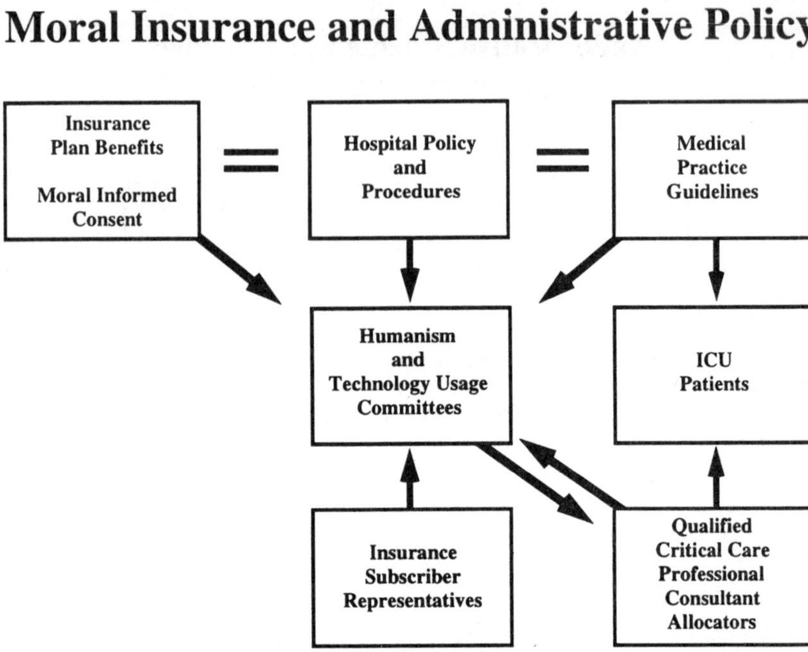

Fig. 3. Health plan morality in Figure 2 employs the Humanism and Technology Usage Committee to function on site in the hospital as a combined source of economic authority and resource allocation oversight function of the critical care consultant gatekeeper.

ogists, pulmonologists, oncologists who will be the principle fiduciary trust physicians within the meaning of the doctor/patient relationship.

In Figure 3, the moral insurance and schematic administrative policy organization is further clarified. The health plan creates and directly funds a committee that is identified as the Humanism and Technology Usage Committee. The committee is composed of representatives from the hospital corporation, the medical staff of primary physicians, critical care consultants, and payors and patient subscribers. This is a legally constituted, "watch dog process committee" that is specifically created and permitted under the insurance laws of the jurisdiction. Attached to this committee are a group of certified Qualified Critical Care Practitioners (QCCP) in the relevant areas of medicine. These physicians and nurses are direct employees of the health care plan. Although they work as integrated members of the hospital and must possess medical and nursing licensure, the health plan prohibits economic or administrative control of these individuals by the hospital and the medical staff as a contingency of hospital and medical staff physician participation in the health plan. These individuals are salaried and are prohibited from fee-for-service medical or nursing practice in this ICU system. These individuals may not rotate to primary care medical status, but must remain as critical care consultants and resource allocation gatekeepers during the entirety of their clinical experience in the applicable hospital. It is expected that these individuals will primarily direct and manage the efficient integrated "case-managed" practice of critical care for critically ill patients. Primary care physicians shall have active participation in the care of their patients in intensive care units but final authority for questionable usage of technology or usage not sanctioned by the health plan will be under the control of the critical care consultant gatekeeper.

To promote predictable, measurable, clinical performance on the part of both primary medical practitioners and critical care consultants, some form of prognostic scoring instrument will be required for health plan participation of providers and the costs of maintaining the surveillance and audit functions will be borne directly by the health insurance plan and not the providers.

It is expected that there will be wide latitude in the day to day practice of medicine both for practitioners and critical care consultants consistent with the unpredictable nature of medical practice. The comprehensive management of critical care patients will have several elements including the choice and number of specialty consultants and the comprehensive attention to the personal needs and desires of patients. The quality and quantity of cognitive service and communication with patients and families will be evaluated by objective scoring systems for their performance. Financial remuneration of both primary care givers and critical care consultants will have a definable component related to customer satisfaction. The terms of these arrangements will be stated within the health insurance plans as arranged by the creators and contractors of the plan. Closed ICUs in this system will be considered to be an arrangement that leads to a conflict of economic interest and will be prohibited from the malpractice tort exemption statute for the jurisdiction as described below.

C) Description and Scope of Activities of the Humanism and Technology Usage Committee

The critical care consultant gatekeeper is an agent of the committee and reports to the committee and its administrative structure. Should the gatekeeper have irreconcilable conflicts with the committee, there would be a secondary level for conflict resolution within the main body of the health insurance plan. If at any time the gatekeeper believes there to be a breach of the contract conditions on the part of the health plan regarding the resource allocation policy to individual patients, he or she would be free to directly notify the insurance authorities without prejudice. The sensitive and vulnerable status of the gatekeeper position will have specific protections in statute.

The Humanism and Technology Usage Committee is a medicolegally binding decision making body expressly recognized under the laws of the jurisdiction. It is the economic authority to dispense or deny funds to the hospital and its medical staff. It also serves as the locus for patient appeals by patients for care rendered in the intensive care unit. It is the locus for denials of individual care occasioned by the entitlement limitation process described below.

The original contract for the health plan describes the appellate process guarantees to patient subscribers. It provides patients and providers with a specifically defined process for resolving grievances from all sides (See Section I below). The rights and process are spelled out in the health plan. Under the laws of the jurisdiction the patient is required to go through the appellate process and the appellate process will be binding upon resource allocation by the health plan. The only exception to this rule will be evidence of a sufficient nature to a court of law indicating that there was a breach of the due-process provisions in the health insurance contract. Should such an appeal be made to a court of law and that appeal found to be without merit by a court, then the costs of all medical care between the time that such appeal was made and the final disposition of the matter will be ascribed to the patient. In the alternate circumstance where a court of law determines that a violation of the due-process provisions of the contract occurred, the plan will be liable for the cost incurred from the time of appeal until judicial determination was made and the plan shall be assessable for a multiple of those costs as a punitive damage payable directly to the patient.

D) Malpractice Laws and Malpractice Provisions

The health insurance laws of the jurisdiction where this contract would be written would be similar to those encompassed under Section 10 of Senate Bill 27 in the original Oregon Basic Health Services Act of 1989. In the Oregon Basic Health Services Act, there is a section prohibiting citizens from bringing malpractice actions in court against health care providers for failing to provide that level of service that was deemed un-fundable by the legislature in the Medicaid benefit package. The legislature of Oregon further granted that the tort exemption would apply to health insurance plans under the employer mandate sections of the law. Thus, an Oregonian managed health care plan would have such exemptions provided they met the minimal basic benefits of the Medicaid program in that state. For purposes of this integrated health insurance morality proposal, the nature of such legislation would apply and a bifurcated standard of care would exist in which providers will still be held to the standard of medical expertise and judgment that exist under the conventional tort law while the standard of resource utilization under the health plan would exempt actions in tort for following the contractually agreed to health plan (Morreim, 1991a; Morreim 1991b; Morreim 1989).

E) Equity in Limitation

Many critical care professionals look at the issue of triage as a situational activity that occurs when there is an acute shortage of beds and an excess of patients seeking access to the beds. This behavior is likely to lead to unpredictable outcomes for individual patients with regard to their entitlement within a health insurance plan. It is likely that legislatures and courts would look upon situational triage as an unpredictable discrimination in the care of individuals given that some individuals might be forced to leave an intensive care unit on a day when there was busy surgery but not at other times when the hospital census was light. A standard of predictability of decision making is imposed on the health plan and situational triage of and by itself is prohibited within this basic benefit health policy.

The abolition of situational triage will be argued to be a destructive event by the prior institutionalization and practices of critical care providers. But that is what is means to function in a resource limited world. The policy must remain coherent and credible not only in the eyes of providers, but in the eyes of patient subscribers who purchase their health insurance in advance to protect themselves against the perils of the natural and social lotteries of life. If people want situational triage and it is important to them, then the health insurance market, being competitive and freely constituted, will make arrangements for it to exist. Thus, people can have additional "rider provisions" to their health insurance that may be acquired at the time the health insurance is purchased. *A la carte* health insurance choice also has a positive social benefit to the least well off class of society. To the extent that more people elect to pay out of their own personal after-tax dollars, additional tax revenue is created that could be targeted for use to increase benefits to the poor.

The abolition of situational triage is also necessary to maintain the predictable nature of justice for critical care outcomes and the specific sanctioning of these rationing systems by both the legislative and judicial branches of government. This system of moral judgment making is one in which the power of contract law takes precedence over tort law (Havighurst, 1986a; Havighurst, 1986b). If providers of health care claim to be professionally encumbered by the extent to which tort law has dominated, controlled, and distorted the distribution of health care benefits to their patients, then they will take up and champion the creation of appropriate health insurance contracts for their patients to facilitate appropriate allocation of resources in a morally contentful public manner.

F) Applying Prognostic Scoring Probabilities of Outcome to the Intensive Care Unit Entitlement Index

The science of outcomes prediction and objective validated scoring systems in critical care is detailed (Kraus, et al.; LeGall, et al., 1984; Le Gall 1993; Lemeshow). This essay does not propose to discuss the limitations of these technologies but simply to state that they exist and their degrees of uncertainty are now quantifiable and publicly recognized. We should acknowledge that there is moral confusion when it comes to the termination of curative care for patients residing in an ICU. Those who oppose the imposition of these external constraints to physicians' autonomy point to the degrees of imprecision that exist in these systems and the degree to which that imprecision may be equal to, inferior, or superior to that of the individual practitioner's clinical judgment. Clinical judgment plays a very important role in validating external systems and no claims are made against it. However, in a system where clinical judgments exercised by individual clinicians are publicly perceived to be in a moral conflict of interest with the allocation of resources, it is necessary for the public welfare to establish a system of independent calibration and validation of those judgments so as to have an economic system for the swift daily and implementable flow of resources for all patients covered in the health insurance program.

G) Clinical Uncertainty, Prognostic Scoring Uncertainty and the Cost of Statistical Certitude

Clinical medicine is not a precise science due to the biologic variability of patients. In the best of diagnostic systems in medicine, there is a degree of statistical imprecision and error. Primary physicians, critical care physicians and nurses, and patients, and families all cling to the American Health Care Ideology especially when their individual care is at hand. Accordingly, there will be no resolution to the cost problem in critical care until we accept reasonable probability estimates of failure and avoid the desire to seek miracles and the 1 in 1,000,000 or 1 in 10,000 chance of success for an individual patient. When a patient resides in an intensive care unit and the probability of mortality is estimated to be 95% or greater at the end of a week, as suggested in the sample policy, there always will be a statistical measure of imprecision around that value. But the value itself chosen by plan negotiators to be the standard for resource limitation is not a medical decision and it is not strictly a financial one. Rather it is the creation of a morality that will determine the degree of imprecision that we as a society or we as a series of contractors in a health insurance plan can afford in what will subsequently be our own personal medical care. Of course, when we are well and healthy and we create the health insurance we will not know whether we will require that category of critical care or whether we will never require critical care. However, that is what it means to create a cost-contained, morally contentful health insurance package within a community of individuals attempting to insure themselves against the perils of the natural and social lotteries of life. The notion that there must be absolute certainty before care is withdrawn, as is so commonly heard every day in the intensive care units of the United States, is simply a delusion that often has been adopted by health care providers and citizens at large as they have absorbed and internalized the American Health Care Ideology. If they wish to continue this ideology, they should have no compunction about further diminishing their access to the other things besides medical care that would make their lives worthwhile. As the cost of health care grows to twenty, thirty, or forty percent of the gross domestic product of the country, people should anticipate that there will be less higher education, less capacity to own one's own home or to afford such amenities as vacations, clothing, or expensive text books of critical care medicine.

Though the latter may appear somewhat harsh on first impression, the younger generation of researchers will soon be making new analyses of the cost of seeking certitude of death prediction. They will measure the actuarial costs for varying probabilities of error for individual patient predictions. As it will take a given quantum of resources to decide that there is objective evidence of mortality at the 95% probability and that the variation of certainty around that will be a given confidence interval, the research is likely to show that it may cost a large multiple of that figure to decrease the confidence interval. It has been the failure of the senior generation of physicians to confront this openly and to share prognoses openly with patients and their families that now are the source of legally contentful suits against physicians for failing to give fatal prognosis information to patients and their families.

I have laid the foundations for the role of health services research in providing the necessary information to put a monetary value on the cost of probability of fatal outcomes. Even in this system, there always will be a small degree of error. There will be an occasional patient, who might have lived had the resources been allocated, but who was denied those resources in the health insurance product. This has been called the technological imperative applied to the principle of rescue so dominant in the medical culture in the latter 20th century. But a morally contentful cost containment strategy must encompass a way to create the limits to these expenditures and to lower the relative priority of these expenditures in comparison with other expenditures that are now conceived by the health insurance plan to hold a higher moral and economic priority. As these packages of health insurance would be freely negotiated during periods of health, one would create the quantifiable definitions of unfortunate vs. unfair outcomes. For those who find such limits to be morally objectionable either on personal or religious grounds, it will be necessary for governments and health insurors to sanction the concept of fiscal accountability for moral diversity outlined in the SCCM Triage Consensus Statement. The Triage Statement clearly delineates that autonomy is not the master but the servant of communal morality in this setting. This is likely to cause major confrontation in the greater political process and will require a maturation on the part of the occidental peoples to realize that individual freedom of patients and their physicians is immoral unless it is circumscribed by the larger morality of the community in which the individual patient and the individual physician reside.

H) Medico Legally Binding Patient and Physician Appellate Process for Entitlement Limitation in the ICU

Limiting situational triage will permit resolution of conflicts in the care of individual patients in the ICU. A health plan offering managed critical care services within the definition of the laws of the jurisdiction in which the health plan operates will require a publicly promulgated list of entitlement screens that identify cutoffs that form the basis for the limits for entitlement to care in an ICU. When we describe these matters in health insurance plans, we are talking about reimbursement only. Should hospitals, physicians, and patients elect to undertake care beyond the limits of the health insurance plan, that would be a matter for them to negotiate privately.

To confront the emotionally charged appeals of patients and their primary physicians, the health insurance plan must be robustly constructed within the meaning of contract law, recognized by legislative authority, and be subject to the sanctions of the judicial system. In this proposal it will be noted in Figure 3 that the central source of economic moral authority in this horizontally integrated health care system is the Humanism and Technology Usage Committee. This is a conceptually new version of a medicolegally binding ethics committee organized to function within a hospital but under the purview of a health insurance plan. This is necessary to eliminate the division of responsibility and authority between insurors and providers in the American economy. This method assures that there will be legal accountability of the health plan as an integrated portion of a health provider system. This would occur, for example, in a large Health Maintenance Organization (HMO), which owns hospitals and has medical and other providers working as its employees. The large HMOs of the United States have been particularly resistant to this conceptual formulation for fear that they would be accused of primarily looking after their own profits. As previously stated, no system of financial incentives is morally adequate to deal with the entire spectrum of individual health care issues and therefore the following structure is proposed.

A committee is formed by the creation of the health insurance plan and has active participation and voting rights by the hospital corporation and those fiduciary trust physicians who will have primary responsibility for the care of patients when they are admitted to critical care units. To assure that physicians and hospitals will be active, equal participants and not dominate the situation, they agree by economic and legal contract to function within the moral limits of health care benefits assigned to subscribers of the health plan as a contingency for their initiation and continued participation in the health plan. Patients and subscribers will have an active role in the creation of the health plan. The health plan would be permitted to have legislatively created tort exemptions as described below unless employers originally sponsoring the health insurance plan had documented an insufficient level of employee participation in the creation of the plan and the on-going assessment of the benefits and restrictions to individual care benefits within the health plan. Employee organizations or labor components of industries would have an active seat on the Humanism and Technology Usage Committee. The Humanism and Technology Usage Committee would have two levels of function. The first level would be the moral creation of health insurance limitations during the contracting process for the creation of critical care benefits. A subcomponent of the Humanism and Technology Usage Committee would be the "Oversight Committee" that would exist for critical care units in a specified hospital. It should be noted that we are dealing only with large hospitals and large critical care components that would meet the criteria for managed critical care services plan. The need for comprehensive case management and the division of tasks between primary fiduciary physicians and Qualified Critical Care Professionals as consultant gatekeepers will preclude easy implementation in smaller hospitals. This essay does not address the issue of efficiency or inefficiency of critical care operations in smaller hospitals. However, there is already documentation of inefficiency of high technology usage in small hospitals relative to the cost of technology and it is unlikely that people in small communities will continue to possess the freedom to squander large health insurance plan resources for that kind of care. A system of regionalization of critical care is already widely acknowledged to be necessary and is actively supported as a matter of health policy by critical care medical organizations like the SCCM.

The Humanism and Technology Usage Committee is a formal structure with a mission, a budget, and contractually defined authority to carry out the individual entitlement limitations of the health care contract. It is a medicolegally binding ethics committee sanctioned in law and health insurance. If a hospital wishes to use this structure to deal with individual ethical issues that do not directly confront resource allocation, the hospital may elect to contract with the health plan to have this function carried out by the Humanism and Technology Usage Committee.

Qualified Critical Care physician Practitioners (QCCP) with specialty certification credentials from any of the existing specialty boards within the American Board of Medical Specialties (or professional certification bodies in other countries) will be qualified to function as critical care consultants in this system. They will be consultants and case managers only for the care of the critically ill. They are prohibited from being primary, fiduciary trust, admitting physicians and the advocates of patients and their rights of consumption within the health plan. To assure that there is no moral economic conflict, these physicians may not become part of the admitting physician component of the hospital medical staff for patients who would require or could reasonably be anticipated to require critical care services. The contractual mechanism of the health plan, hospital corporation, and the hospital medical staff component would define the limits of abuse for this privilege and detail the grounds for termination of thc critical carc physician from health insurance plan employment.

Primary fiduciary trust physicians who would act as the advocates of their patients would be entitled to freely and robustly advocate for the individual circumstances of their patients in a dialogue with the critical care consultant and members of the managed critical care services team. Prior to becoming a recompensable, qualified, admitting physician for patients in this critical care system, the primary fiduciary trust physician will have read and understood the moral economic limits to health benefits that would likely become implementable in the specialties in which he/she works. While some of these limits would obtain directly for issues related to prognostic scoring, it is anticipated that others would relate to the reasonableness or unreasonableness of undertaking medical or surgical, high technology strategies prior to admission to an ICU that would subsequently make ICU care and physiologic support necessary. In this essay we look only in this initial conceptualization of prognostic scoring in defining the moral limits of individual entitlement of patients to consume resources and the autonomy prerogatives of their physician advocates to request such services and requisition them from a health insurance plan.

From time to time, a patient will be identified (through computerized entitlement limitation resource consumption screens) whose limit has been exceeded. At such point automatic notice of this circumstance would be communicated formally to the primary physician advocate and the patient or his legally documented surrogate. An appropriate period of time (within defined limits) for independent personal response of the patient and the physician advocate would then be permitted. There would be no discretion on the part of the critical care physician consultant gatekeeper as to what the screens would be, as they already would have been decided in the health insurance plan policy. The authority of the critical care gatekeeper is circumscribed in terms of notification of entitlement limitation excess by the rules of the health plan. He/she functions and reports to the Humanism and Technology Usage Committee exclusively for all matters that have to do with resource allocation regarding the entitlement limitations in health insurance. As the original model (given as an example only) already contains the definition of medical futility, it will be difficult for individual patients or physicians to make appeals that would have standing in the health insurance plan. The appellate process is designed to assure that all of the necessary information gathered by the critical care gatekeeper and the Humanism and Technology Usage Committee is factually correct. When that discovery process has been exhausted, a decision will be made (usually by the critical care gatekeeper) and, if contested, then by the Humanism and Technology Usage Committee. In this system, the committee's decision is final after the individual discovery appellate process has been concluded and all funding for critical care will be terminated. However, for those patients who do not die or are not anticipated to die with the removal of active critical care therapy, the health plan will pay for and assure that the necessary services for critical care hospice services or intermediate care will be fully funded by the health plan. Once a patient enters that category, comfort measures and supportive care will be provided as will counseling and other necessary humanistic supports, and will be supervised by the Humanism and Technology Usage Committee.

Physicians attempting to reestablish active therapeutic treatment in such patients may not do so without obtaining grounds for such that would be acceptable to the Humanism and Technology Usage Committee. Any unilateral decisions taken by primary, fiduciary trust physicians with or without the active or passive permission of the hospital will put the patient, physician and hospital at financial risk.

If termination of entitlement and triage out of the intensive care unit is carried out under the purview of this policy and is contested after the discovery appellate process has run its course, then a formal hearing before the Humanism and Technology Usage Committee will occur and minutes of the proceedings will be kept. The final review will be an impartial review of the gatekeeper's report to the committee with patient's surrogate and primary attending physician invited to attend. Under such circumstances, we are working under the legal doctrine of contract law. It was specifically noted that as a contingency for entering the health insurance plan both the patient and the physician were fully informed and knew the policies and procedures that would be applied should patients come for care in an intensive care unit. Any notions of ignorance or not understanding these things will not be acceptable in a court of law because there will be rigid standards required to acquire the health insurance and to enter into the plan either as a patient or as provider in the original instance. Courts would likely rule in favor of the health plans because the necessary legislative authority would have defined the policy requirements demanded by the judiciary.

To allow health plans to function across several jurisdictions, or in large interstate health care enterprises, federal legislation will have been created that will prohibit civil litigation in either state or federal courts against any physician or nurse for adhering to the triage policies created by the contract process for acquiring moral authority in health insurance. Medical negligence suits will continue to exist but will be limited by the absolute resource standard of availability in Section 10 of Senate Bill 27 of the Oregon Basic Health Services Act of 1989. The fact that a person would die more swiftly than they would have by the continued administration of high technology care is not a necessary and sufficient grounds legally for seeking recompense and is legally recognized as a morally and legally unfortunate but not unfair circumstance within the health insurance system.

It will be noted from Figure 3 that physician practice guidelines are mentioned as opposed to physicians while hospital policies and procedures are mentioned in contradistinction to a hospital corporation. The reason for this formulation in this conceptual health insurance plan is that there must be latitude accorded to both institutions and to individual providers in the creation of these policies so that they may freely advocate for patients in appellate processes. It is anticipated that as such a system enters implementation within the Oregon Health Plan, several mid-course corrections in hospital policies and practice guidelines, as well as, the entitlement limitations within the health plan will be made by virtue of economically mandated dialogue between the parties most affected by these issues. This is entirely appropriate as the nature of political and economic conflict is best resolved within the health insurance plan.

Patients like Helga Wanglie, who hold specific religious or moral views concerning the sanctity of life in the vegetative state, could elect to buy "right to life premium surcharge insurance" when they enter the insurance plan. All systems that choose to forbid a behavior on moral grounds are in trouble. They indicate a lack of tolerance for the freedom of people to hold religious and moral views within a pluralist society. As the American society openly supports the free exercise of religion, we should respect those who hold religious views as they would affect the allocation of health care resources. But when those views require the use of resources that would act to limit the care of others in a resource-constrained society, then the health plan must specifically acknowledge that they can no longer be, free riders in a resource-constrained society.

Objections to this model will be many. The creation of the insurance product as it was in Oregon will be difficult and time consuming. Progress in medicine, science, and pharmacology will require continuous updating about what will count as acceptable limitations as the boundary zones of medicine will be changing. As new technologies are developed, entitlements may change up or down depending on how the cost of new technologies will limit the capacity of the system to offer the same entitlement structure in one year as it was in the prior. Examples of items in this category would be a large series of biotechnology engineered molecules for the treatment of septic shock that will be patented and becoming available in the marketplace that will place enormous economic demands for the purchase of drugs for critically ill people. As more of these molecules become available there will be a need for technology assessment as well as a need to delay the use of molecules where the public and physicians will clamor for them. The extent to which these penetrate the marketplace of general health care costs will impact on the capacity of the health care system in the ICU to offer more traditional and less costly treatments to other patients within the same environment.

SAMPLE POLICY OF EXPLICIT HEALTH CONTRACT

"This policy limits payments for futile critical care. For the purposes of this contract futile care is defined by the signatories to be a prognostic prediction of death of 95% probability after 7 days of ICU care. It is recognized that patients may wish to have more or less degrees of probability attached to the determination of this finding. Accordingly, this contract offers a series of probability tolerances to the certitude of the ultimate diagnosis of futility. The basic package of ICU benefits in this health plan carries a probability variance of X%. Should you decide as a purchaser of this health plan that you would like to have a higher level of certainty attached to these outcomes, you may elect these degrees of probability as expressed in the premium surcharge table defined by our actuaries in the Appendix of this health insurance policy. These option packages are available annually at the renewal of the contract and may not be changed during a covered benefit year".

CONCLUSION

There will always be a divergence of opinion concerning the proper public policy regarding the provision of ICU resources. To the extent that we are discussing the investment of tax resources, communal resources, or resources garnered through the group purchase of a health insurance plan, there will be a need to arrive at a common morality that will control the disbursement of funds to the care of specific individuals as a moral matter in health insurance policy. There has been great resistance to the notion of "economic morality" not only in theological groups but among secular bio-ethicists as well. The arrival of the Oregon proposal represents the coming of a new Martin Luther to Christian conceptions of what constitutes virtuous health care in the 21st Century. The task at hand is to lay the foundations for a reconceptualization of Christian mores in health insurance. Perhaps the greatest challenge will be the simple articulation that there is religious grace that could be achieved by promoting the infusion of prognosis based criteria into the allocation of scarce resources at the margins of biologic life in ICU.

NOTE

[1] For a better understanding of this debate see *Health Affairs*, Fall 1991, Intenational Health Spending Comparisons, David A. Rublee and Markus Schneider with rebuttal by Scheiber and Poullier, pp. 1987–2011 and *The Journal of Medicine and Philosophy* 18, October 1993, Methodological and Conceptual Issues in Health Care System Comparisons: Canada, Norway, and the United States, Baruch A. Brody and Reider K. Lie with responses by Poullier.

REFERENCES

Ashby JL, Jr., Craig KL: 1992, 'Why Do Hospital Costs Continue to Increase?', *Health Affairs* 11(2), 134–147.

Battin M: 1982 *Ethical Issues in Suicide*, Prentice Hall, Engelwood Cliffs, New Jersey.

Berenson R: 1984, *Intensive Care Units (ICUs): Clinical Outcomes Cost and Decision Making: Health Technology Case Study, No. 28.* Office of Technology Assessment, Government Printing Office, Washington, D.C.

Cruzan vs Director, Missouri Department of Health: 1990, U.S. Supreme Court No. 88–1503. The United States Law Week 58 LW 4916–4941.

Cullen DJ, Civetta J, Briggs BA: 1974, 'Therapeutic Intervention Scoring System: A Method for Quantitative Comparison of Patient Care', *Crit Care Med* 2, 57–60.

Ethics Committee of the SCCM: 1994, 'Consensus Statement on the Triage of Critically Ill Patients', *Journal of the American Medical Association* 271, 1200–1203, (See Appendicies).

Engelhardt HT, Jr., Rie MA: 1986, 'Intensive Care Units Scarce Resources and Conflicting Principles of Justice', *JAMA* 255, 1159–1164.

Feshbach M: 1988, 'Health in the USSR – Organization, Trends and Ethics', in Sass HM, Massey RU, (eds.) *Health Care Systems*, Reidel Publishers, Dordrecht.

Havighurst C: 1986a, 'Altering the Applicable Standard of Care', *Law and Contemporary Problems* 49, 265.

Havighurst C: 1986b, 'The Changing Locus of Decision Making in the Health Care Sector', *Journal of Health Politics, Policy and Law* 11, 697.

In Re Quinlan: 1976, 70 NJ 10, 355A 2nd 647 (Cert) Denied Sub Nom Garger vs New Jersey 429 US 92297 SCt 31 9 50L Ed 2nd 289.

Keene AR, Cullen DJ: 1983, 'Therapeutic Intervention Scoring Systems: Update 1983', *Crit Care Med* 11, 1–3.

Knaus WA, Wagner DP, Draper EA, *et al.*: 1991, 'The Apache III Prognostic System: Risk Predition of Hospital Mortality for Critically Ill Hospitalized Adults', *Chest* 100, 1619–1636.

Knaus WA, Draper EA, Wagner DP, *et al.*: 1986, 'An Evaluation of Outcome from Intensive Care in Major Medical Centers', *Annals Int Med* 104, 410.

LeGall JR, Loriat P, Alperovitch A, *et al.*: 1984, 'Simplified Acute Physiology Score for ICU Patients', *Crit Care Med* 12, 975–997.

LeGall JR, Lemeshow S, Saulnier F: 1993, 'A New Simplified Acute Physiology Score (SAPS II) Based on a European/North American Multicenter Study', *JAMA* 270(24), 2957–2963.

Lemeshow S, Teres D, Klar J, *et al.*: 1993, 'MPM-II: Mortality Probability Models Based on an International Cohort of Intensive Care Unit Patients', *JAMA* 270, 2478–2486.

Morreim EH: 1991a, *Balancing Act: The New Medical Ethics of Medicine's New Economics*. Kluwer Academic Publishers, Boston.

Morreim EH: 1991b, 'Justice and Health Care Rationing: Lessons from Oregon', in Strosberg MA, Wiener JM, Baker R, (eds.). *Rationing America's Medical Care: The Oregon Plan and Beyond*, Brookings Institution Press, Washington, DC.

Morreim EH: 1989, 'Stratified Scarcity: Redefining the Standard of Care', *Law, Medicine and Health Care* 17, 356.

Murphy DJ: 1988, 'Do Not Resuscitate Orders: Time for Reappraisal in Long Term Institutions', *JAMA* 260, 2101.

Office of Technology Assessment Task Force: 1987, *Life Sustaining Technologies in the Elderly*, Washington, D.C. Government Printing Office.

Oregon Health Services Commission: 1991, *A Report to the Governor and Legislature: Prioritization of Health Services*, Salem, Oregon, pp 1–75.

Pellegrino E: 1992, 'Truth Telling to the Patient: A Cultural Artifact?', *JAMA* 268, 1734–1735.

Pontoppidan H, Wilson RS, Rie MA, *et al.*: 1977, 'Respiratory Intensive Care', *Anesthesiology* 47, 96–116.

Rapoport J, Teres D, Lemeshow S: 1994, 'A Method of Assessing Clinical Performance and Cost Effectiveness of Intensive Care Units', *Critical Care Medicine*, 22, 1385–1391.

Rie MA: 1989a, 'A Social Responsibility to Die? Freedom, Aging and AIDS', *J Clin Anesth* 1, 222–227.

Rie MA: 1989b, 'Professional Ethics and Political Power', *Rationing of Medical Care for the Critically Ill*. Edited by Martin A. Strosberg, I. Allen Fein, and James D. Carroll, Brookings Institution Press, Washington, D.C., pp 82–86.

Rie MA: 1989c, 'Regionalized Intensive Care Units and Medicare: Ethics, Public Policy and Outcome Research', *Crit Care Med*, 17, 704–705.

Rie MA: 1992, 'Practicing Medicine, Fiduciary Trust Privacy, and Public Moral Interloping after *Cruzan*', *The Journal of Medicine and Philosophy*, 17, 647–664.

Rie MA: 1992, 'A Further Comment on Acquired Immunodeficiency Syndrome and AIDS', *Crit Care Med*, 20(3), 320.

Russell LB: 1982, 'The Role of Technology Assessment in Cost Control', in McNeil BJ and Cravalho EG (Eds.): *Critical Issues in Medical Technology*, Auburn House, Boston, Mass, pp 129–138.

Shabot MM, Bjerke HS, LoBue M: 1992 'Quality Assurance and Utilization Assessment: The Major by Products of a Clinical Information System', in *Proceedings of the XV Symposium on Computer Applications in Medicine*. American Medical Informatics Association, Bethesda, Maryland, SCAMC 1992, 15, 554–559.

Schieber GJ, Poullier JP: 1989, 'International Health Care Spending', *Health Affairs*, 5, 112–122.

Schneiderman LJ, Jecker NS, Jonson AR: 'Medical Futility: Its Meaning and Ethical Implications', *Annals Int Med*, 112(12), 949–954.

Shortell SM, Reinhardt UE: 1992, *Improving Health Policy and Management: Nine Critical Research Issues for the 1990's*, Health Administration Press, Ann Arbor, Michigan, 1992.

Smedira NG, Evans BH, Grais LS, *et al.*: 1990, 'Withholding and Withdrawal of Life Support from the Critically Ill', *New Eng J Med*, 322, 309–315.

Surbone A: 1992, 'Truth Telling to the Patient: Letter from Italy', *JAMA*, 268, 1661–1662.

Tomlinson T, Brody H: 'Futility and the Ethics of Resuscitation', *JAMA*, 264(10), 1277.

The Sacred Congregation for the Doctrine of the Faith. *Declaration on Euthanasia* Vatican City, May 5, 1980.

Youngner SJ: 1988, 'Who decides futility?', *JAMA*, 260, 2094–2095.

MIGUEL A. SANCHEZ-GONZALEZ

COMMENTARY ON: BRODY & RIE

I

The papers by Baruch Brody and Michael Rie focus on the criteria and procedures for admission, maintenance and discharge of individual patients from intensive care units.

Although they differ somewhat in their approach to allocating scarce resources, they share common ideas and proposals:

1. They give high weighting to the severity of illness scales in the domain of individual decision making. For Brody it is only the patient who can make these judgments. For Rie, such measurements can be imposed in the clinical practice to the extent that they are used to calculate an *entitlement index to ICU care* which can be a kind of Ulysses contract. In either case, to give a decisive weight to probability in life and death individual decision making, is to adopt a point of view that seems to depart from a principle of the sanctity of life.
2. Both assume the necessity of rationing the resources to the individual patients. Brody insists that a physician must acknowledge a web of social loyalties besides a devotion to his or her present patient. While Rie promotes the infusion of prognosis based criteria into the allocation of scarce resources, assuming therefore a utilitarian position.

Nevertheless, the main difference between the two authors revolves around the question of a physician's role in rationing resources. Brody argues the necessity of investing the practicing physicians with a rationing and cost containment power. While Rie argues for a separation of the two functions of allocating resources and advocating the interests of the individual patient. He maintains that this is crucial for the fiduciary trust relationship between the physician and his patient.

Brody's paper attempts to clarify a variety of issues by dividing them into two major sections. In one section, he deals with the criteria for admission and discharge, when the question is just whether or not

K. Wm. Wildes (ed.), Critical Choices and Critical Care, 57–61.

that is beneficial to the patient. The other section refers to the further complexities that arise, when economic considerations play a role either in the triaging context or in the cost-containment context.

As for the problem of what should be considered, in each case, as medically beneficial care, Brody proposes that the physician ought to inform the patient about his objective probabilities of survival, based upon severity of illness measurements, in order to allow the patient to make his own treatment decisions. Even if it is yet to be known to what degree that information will induce patients and their families to forego ICU care. Probably, most patients will want intensive care even with a modest life expectancy.

Brody also wonders whether there are some cases of futility where the physician's informational role changes to that of decision maker. He correctly sees that the idea of futility is always value laden and, consequently, he does not see how can the patient's values be overcome without having some objective and superior theory of the Good. In that sense, to accept the physicians' futility arguments is also, to accept their value systems of what is not worth further treatment.

When discussing resource allocations, Brody's main point is that the practicing physician should exert a rationing function. For not to do that rationing, as he says, and continue with a single-minded devotion to the individual patient would lead to injustices in the allocation of the public funds. Moreover, he sees the physician as not being morally free to adopt the traditional approach of advocating exclusively his patient's interests. The physician, like every other person, lives in a web of loyalties, each of which make demands upon him. These demands can be conflicting. Nonetheless they must be balanced in every case.

Although Brody defends a rationing role for the physician and gives good arguments for it, he does not deny that the physician must remain loyal to the patient. What is not so clear is how can this be implemented. How can the conflicts be solved? We may even wonder whether this double function is actually feasible. Can a servant serve two masters?

As a conclusion, Brody stresses the wide variety of ICU patients and interventions. He stresses the necessity of specifying which interventions are to be considered for which patients. While paying great attention to the clinical details of different cases, he ends by opening the way to a kind of ICU casuistry and reminds us about the necessity of recognizing the undeniable benefits as well as the cost of that casuistry.

II

Rie's paper states the need of a new economic morality, as "a publicly articulated and value content laden morality that would be acceptable to members of secular pluralist societies". In conjunction with that, he proposes a system of assigning ICU resources to individual patients that would prevent conflict of loyalties among individual physicians.

Rie is correct in thinking, that there are few specific guidelines developed in health policy to indicate what the limits of individual entitlement might be. He believes that a major revolution in moral thinking in this area has been articulated in the Oregon plan, which prioritizes the universal health care services and specifies the limits of individual entitlement, based on a kind of moral contract, for the allocation of resources.

When the level of public expenditure is the only limiting factor to acquire technology and labor, as is usually the case, ICUs are always under pressure to make choices of patients and to function in the triage mode. The problem is that moral values are excluded from explicit dialogue, in these systems that are forced *de facto* to ration in a way not always beneficial for a number of patients. Egalitarian systems that provide global budget caps leaving physicians and hospitals to create *ad hoc* morality, would be immoral, in failing to provide a vision for ICU entitlement limitation. To have no explicit criteria would be to open the way to mediocrity and bad assistance and would have worse consequences.

Rie also believes that, behind the actual health policies, is what he calls "the tragedy of morality in Western medicine, where the preservation of life and the sanctity of life has been infused to such an extent into our Judeo-Christian based health care systems, that it risks to destroy the other qualities of existence that make survival worthwhile". Futile care would amount to social injustice. What is needed would be nothing less than a reconceptualization of certain Judeo-Christian values.

Rie is well aware of the fact that developing a morality of disentitlement can bring charges of discrimination against those with disabilities, as it has been the case with the Oregon plan. It can even raise, what shall be termed "the Nazi presumption", leading to dictatorships and extermination of the helpless. Nevertheless, Rie states the need to overcome those charges, finding ways to re-articulate entitlements and disentitlements in a morally acceptable way.

In a previous paper, Engelhardt and Rie have developed an "Intensive Care Unit Entitlement Index" (ICU-EI) to create a rational objective system of patient prioritization (Engelhardt and Rie). In this paper, Rie presents a horizontally integrated system based on an entitlement index, where the ability of patients to benefit from the resource allocation, is the pre-eminent criterion for allocation. In this system, the necessity of preserving the patient trust in the physician and the need for specific gate keeping of the resources, are satisfied by maintaining the two functions clearly separated.

In his conclusion, Rie affirms the need to arrive at a common economic morality, to control the disbursement of funds and strongly supports the infusion of prognosis based criteria into the allocation of scarce resources, at the margins of biologic life in ICU.

III

I shall make some final remarks and point out some issues quite worth discussing.

In the first place, do we all agree that health resources are so scarce that strong measures of rationing are inevitable? Are we no longer allowed to be confident in rationing avoiding strategies? Let us not forget that there are some authors who think that: "Rationing may not be as inevitable as generally supposed and before we can impose rationing on morally valid grounds, all other means of cost containment must be exhausted"(Pellegrino). Still others believe that "The ethical problem of how to apportion limited resources has been forced on us by arbitrary limitation of health expenditure" (Rawls).

Secondly, we can discuss whether some of these rationing proposals may or should lead us away from sanctity of life positions. In this rationing, how and to what extent ought we to take into account quality of life considerations?

Other objections to be discussed are those that have been frequently raised against the utilitarianism in general, and the Oregon's plan for Health Care Rationing in particular (Daniels; Hadorn; Dougherty).

Still a fundamental question remains. Would a system of rationing heath care based on need, on degree of suffering, on rights, or on any other principle of justice, not be more equitable and just than a system that mainly uses prognosis based criteria? How can we circumvent all those considerations of justice?

Notwithstanding, these brave and thought provoking articles raise inescapable issues in clear and well thought out ways. They will be of great help in clarifying our ideas in this delicate field.

REFERENCES

Daniels, N.: 1991, 'Is the Oregon rationing plan fair?', *JAMA*, 265, 2232–2235.

Dougherty, C.: 1991, 'Setting health care priorities, Oregon's next steps', *Hastings Center Report*, 21(3), 1–10.

Engelhardt, H.T. (Jr.), Rie, M.A.: 1986, 'Intensive care units, scarce resources, and conflicting principles of justice', *JAMA*, 255, 1159–1164.

Hadorn, D.C.: 1991, 'The Oregon priority-setting exercise: Quality of life and public policy', *Hastings Center Report*, 21(3), 11–16.

Pellegrino, E.D.: 1986, 'Rationing health care: The ethics of medical gate keeping', *The Journal of Contemporary Health Law and Policy*, 2, 23–45.

Rawles, J.: 1989, 'Castigating QALYs', *Journal of Medical Ethics*, 1989, 15, 143–147.

ALEJANDRO SERANI-MERLO AND PEDRO ROSSO

INTENSIVE CARE ITS NATURE AND PROBLEMS

I. INTRODUCTION

"Intensive care" or "critical care" is a relatively new medical development. Because of its special characteristics and high costs, we tend to consider intensive care as a different kind of medicine. Nevertheless, the basic problems and the type of decisions doctors confront in this area have been always part of medical practice, especially during exceptional circumstances such as war or natural catastrophes. Perhaps, what gives to intensive care its special connotation is the fact that what used to be exceptional in traditional medicine, has become part of the daily routine at intensive care units (ICU's). Furthermore we feel that the complexity of medical practice in an intensive care environment, the physical and emotional strains imposed on the staff, and the relevance of the individual and social elements related to intensive care, deserve a greater attention from both the medical community and the community at large[1]. In this review of certain aspects of intensive care we will attempt to: 1) characterize its technical and anthropological dimensions; and 2) define the nature of decision-making involved in this kind of care. In addition, we will propose actions aimed at improving the quality of the decision-making process, thus helping those involved.

II. WHAT IS INTENSIVE CARE?

1. *Descriptive Approach*

Intensive care units are a by-product of highly technological health care systems. For that reason intensive care units (ICU's) are mostly located in large hospitals. Presently, the most common types of ICU's are: medical intensive care units, surgical recovery units; and neonatal, pediatric, coronary, respiratory and trauma units.

The creation of an intensive care unit requires several elements, including demand for services (patients), a competent medical and nursing staff, appropriate facilities and, lastly, political will. All these elements will be briefly analyzed in the following paragraphs.

K. Wm. Wildes (ed.), Critical Choices and Critical Care, 63–78.

A. Patients

Patients admitted to ICU's suffer from a wide range of medical conditions, some as different as a very low birth weight newborn and a brain dead adult. Despite this wide variety of medical situations patients requiring intensive care fall in two categories: 1) those in *actual high risk of inminent death*,(i.e. critically ill patients such as diabetic coma, *status epilepticus*, respiratory failure or healthy very low birth-weight newborns); and 2) those *potentially at high risk of inminent death* (i.e. patients admitted to an ICU in order to perform a potentially dangerous procedure, such as plasmapheresis or corticoid pulses in a myasthenic patient; provocation tests in some kinds of periodic paralysis; special management of cardiac arrythmias and others).

This proposed characterization of ICU patients includes at least two relativistic elements: *risk* and *inminence of death*. Depending on the criteria on which the terms 'risk' and 'inminent' are based upon, practically every hospitalized patient could be eligible for intensive care. This theoretical prediction has been empirically documented in a study by Strauss and his colleagues (Strauss et al. 1986) where they found an inverse relationship between the number of beds available in an ICU, and the severity of the patients who were admitted.

B. The Personnel

In general, ICU's staff are highly trained and selected by their unique professional and technical skills. ICU personnel are among the most highly trained physicians, surgeons, nurses and physical therapists. Similarly, comparative socio-medical studies on the humanistic perceptions and sensibilities of this personnel are lacking.

The ICU staff works in close contact with their patients and professional/ patients ratios of 1:2, and even 1:1, are not uncommon. Despite this small ratios, since in ICU's medical care is provided by teams, working in shifts that may change their scheduled periods in succesive days and weeks, the patient-physician or family-physician relationship is usually not restricted to one physician or to one nurse. Even when special efforts are made in order to personalize this type of relationships, inevitably, the responsibility over one individual patient tends to be spread among teams of health care workers. Moreover, most decisions, especially the most critical ones, are usually discussed by the team as a whole, a fact that contributes to depersonalize the patient-doctor relationship. This is a critical and particularly sensible point that must be recognized during the decision-making process and when decisions must be put into practice. As discussed below, this problem is compounded by the fact that the original attending physician lost real tuition over the patient when he was transfered or admitted into the Intensive Care Units.

Under usual circunstances, medical care in ICU's is provided under physically and emotionally demanding conditions. The level of stress that this situation creates depends on a plurality of factors such as frustration due to the incapacity to revert a physiological failure; the objective difficulty of some procedures or the time constraints for their application; the pain and suffering of the patients; tension and open conflicts with patient's relatives; fear of legal suits; contrasting opinions or character incompatibilities within the team or between different teams, etc.

C. Medical Technology

Medical care in ICU's relies very heavily on a variety of sophisticated equipment that can be tentatively classified as follows:

1) *Monitoring devices:* These include equipment such as oscilloscopes, respiratory gas analyzers, polygraphs, etc. Monitoring involves a periodic assessment of changes in several internally regulated or medically controllable physiological variables[2].

2) *Diagnostic devices:* This category includes all ultrasonographic, x-ray, radionuclear, or nuclear-electromagnetic resonance imaging equipment. This type of diagnostic equipment is mostly used to establish the presence, nature and location of a given abnormality, but also it can help to evaluate the course of a disease or the results of a medical procedure.

3) *Life-sustaining devices:* This category includes mechanical ventilators, dialysis machines, complete parenteral nutrition equipment, intravenous infusion pumps, temporary pacemakers, etc. These equipments and procedures are used to sustain the failure of vital functions. Following withdrawal of these equipments or suspension of these procedures death would presumably ensue in a variable, but relatively brief, period of time.

Most of the equipment and procedures used in ICUs is also used in other services of any large medical center, but their simultaneous presence and almost uninterrumped utilization is unique to intensive care.

2. *Political Will*[3]

The decision of having an ICU cannot be only justified on the basis of the availability of eligible patients and suitable personnel and facilities. Nevertheless, well-defined criteria concerning the basis for this decision are lacking. From the physician's point of view financial considerations are a rather secondary aspect. If the hospital is technically qualified to open an ICU, the institution often feels the ethical imperative to offer that kind of help to the patients. If, however, despite the perceived need and feasibility of creating an ICU, the administrative authority decides that financial resources cannot be allocated, the individual members of the hospital are not morally responsible toward the commnunity for not providing that kind of assistance[4].

For the medical team the problems begins when resources *are allocated* since when the ICU is operating, the capacity to assist all the patients in need will be always finite. On this respect it is important to remember that, as previously mentioned, in principle, all hospital patients could be eligible for intensive care.

3. *A Proposed Definition*

Based on the previous considerations, intensive care can be defined as medical care provided to patients at an actual or potential high risk of inminent death by especially trained and exclusively dedicated personnel who actively and continuously monitorize critical morpho-physiological variables and can rapidly intervene to avoid physiological breakdown. Once an ICU is in operation – following a political decision – it requires an implicit or explicit criteria for patient selection assistance.

III. SOME ASPECTS OF DECISION-MAKING IN INTENSIVE CARE

Among the great number of important decision taken every day in ICUs two type of decisions are particularly important, namely: 1) Eligibility for intensive care and 2) Course of action in patients admitted to an ICU.

1. Who Is Eligible for Intensive Care?

The ICU staff is usually requested to admit in their unit patients referred by other physicians. For this reason we feel that to a large extent the ethical responsibility of determining eligibility for intensive care rests with the physicians requesting the admission.

Ideally, the attending physician should request admission in the ICU for patients in need of for critical care (see below) and who are willing to receive intensive care. In many cases the patients themselves are unable to participate in this decision because of their general condition. Establishing if an individual patient is willing to receive intensive care is an aspect even more difficult to deal with. In fact, determining the validity of the perceived medical need – as difficult and uncertain as it is – is only one of the elements that the patient must take into consideration when determining if he will consent to receive intensive care.

Additionally, every ICU should have criteria concerning the role of the attending physician after the patient is admitted. In our experience this is neither clearly nor uniformly determined.

Since the medical need for ICU care can hardly – if ever – be determined on absolute grounds, the decision to admit a patient must be based on conjectural or probabilistic considerations. Here is where criteria such as the APACHE III prognostic system (Knaus et al., 1991) or the prognostic criteria for anoxic coma can help, (Shewmon & De Giorgio, 1989)[5]. In a recent review of this topic, Kilner (1990) has identified five different kinds of medical criteria for selecting patients for health care: medical benefit, inminent death, likelihood of benefit, length of benefit and quality of benefit. After analyzing their pros and cons, Kilner concludes that only the first three can be retained as ethically acceptable.

The question concerning the right of an individual patient to receive intensive care is a matter of social justice that will not be adressed here. Kilner (1990) examining the four social criteria that he could identify retains as valid only those based on the amount of resources required and the one based in the special responsibilities that the patient has. We believe that the decision of using intensive care should not rest solely in the individual doctors and the patients or their relatives. Each hospital should have a set of rules concerning criteria for admission and discharge from ICUs. Besides relieving the physicians from the burden of assuming that type of delicate decisions all by himself, those type of rules have obvious practical implications including avoiding unnecessary admittances or long stays influenced by pressures other than medical needs. We believe that admission and discharge rules should include the following points. First, clear, explicit and public rules about what kind of patients *will not* be eligible for intensive care. Second, explicit criteria concerning the medical benefits that a patient should expect from intensive care. These criteria must be widely known and in current use by the attending physicians entitled to request admission of their patients in the ICU[6]. Third, a demand for proficiency in clinical ethics for the attending staff, and of attending physicians. Fourth, explicit criteria on the role of the attending physician's directives in relation to the type of intensive care that his patient will receive once admitted in the ICU.

Scholars and health politicians have payed little attention to the role of individual physicians as real and final resource "allocators". We are not aware of studies on the budgetary impact and patient satisfaction of different 'styles' of medical care in relation to intensive care. In our experience, the requirements presented above are determinant of the "ethical" quality of medical services offered by ICU's. Quality that in formal ethical grounds can be expressed as a middle term between irrational agressiveness and negligent passivity.

If there are vacant beds and a well justified request for intensive care is presented by an attending physician, it is hard to imagine a reason – on medical grounds – that could justify a rejection to receive that patient. A real ethical problem emerges, however, when the ICU is crowded. As proposed by Kilner (1990, pp.6–7), this situation has a striking parallelism with the classic moral puzzles about shipwrecks and scarce lifeboats. For example, a new survivor is found in the sea and there is only a single, overcrowded, boat. In our ICU puzzle the situation is somewhat different because the captain and the sailors (the ICU team) are not on the boat; thus, they are not eligible to be discharged. To make a closer parallelism with the shipwreck situation, we can imagine that some of the passengers in the boat are injured, and they will surely, or most probably, die before reaching the shore, or if they do reach shore probably they will die shortly. There is also the possibility that a ship carrying medical assistance could find the boat before reaching the shore. What to do? Should the captain throw over board an injured old man to make room for an apparently healthy boy floating nearby? Perhaps the boy will die and the old man would have survived. Who knows?

We do not have an answer to these questions. Moreover, regardless of his decision few of us would indicte the captain. However, if shipwrecks become a daily occurrence, the community must provide captains with certain rules or face the choice of arbitrariness of the captains or to renounce to navigation. So far, no communities have set explicit rules of behavior for their intensive care teams.

2. *What to Do with Patients in the ICU?*

As alluded to earlier difficult decisions are part of the daily routine of ICUs. These decisions are similar in nature to those that physicians must face in their medical practice. However, the frequency and the seriousness of ethical dilemmas confronting ICUs physicians makes necessary to establish a regular decision-making process in order to help the staff and to insure that the patient's rights and dignity are protected. The decision making process that we have established in our teaching hospital[7] is based on the fact that medical decisions and actions always have both a technical and an ethical dimension[8]. A physician faces a technical dilemma when something interferes with the possibility of achieving medical goals or what he considers to be the medical goals. The same physician will face an ethical dilemma when something interferes with the possibility of achieving what he considers to be the goals of any individual person[9]. The technical problems are a question of deciding on the best means to obtain a certain goal – in a situation where the achievement is not questioned. By contrast, ethical problems arise as a question about the goals to be achieved by a given mean. Thus, while the technical dilemma has to do with the search of means in face of unquestioned ends, ethics deals with questions about the ends which are pursued by certain means. Since medicine is only one of the multiple domains in which human beings operate in order to achieve their ultimate goals in life, it is clear that there is a subordination of medicine to ethics and that each medical action necessarily has an implicit ethical dimension, inasmuch as the goals of medical actions are but a part of a person's ultimate goals as an individual.

In intensive care, many medical decisions are uncertain from both ethical and technical points of view. For this reason, each time we confront a clinical dilemma we must try to answer the following questions: What decision appears to be problematic? What kind of problem am I really facing with? Who must decide?

Decisions have to do with actions (internal or external), therefore, in order to clarify the origin of the difficulty we suggest the ICU team to make a list of all the possible actions that surely or very probably could be taken in the near future in a given patient. For example, should the patient be connected to a ventilator if he develops respiratory failure? Should we try to ressuscitate him, if he suffers a cardiac arrest? Should we use total parenteral nutrition? In order to pose these questions the various technical, psychological, socio-economic and personal factors related to the individual patient should be well known to all the members of the ICU team.

As a second step, team members are requested to identify those actions where opinions concerning their appropiateness are divided. For example, the need or advantages of using ventilator or ressucitation maneuvres or the role of parenteral nutrition. After this is accomplished, since the problems posed by patients are not purely technical or ethical in nature, the team is asked to identify the source(s) of their doubts for each alternative considered. The source can be a technical problem (i.e. uncertainty of prognosis or of side effects), or primarily an ethical one (i.e. proportionality of treatment for a known outcome).

For each of the technical and the ethical uncertainties identified, an effort must be made to ensure that all the evidence necessary to reach a decision is available. Usually in this stage of decision-making the opinion of other experts or a review of the literature can be important to clarify the implications of a technical uncertainty. For example, should a biopsy be performed to determine the nature of the underlying pulmonary complication? Is the risk/benefit ratio low enough to justify the procedure? Should we request the opinion of the best known expert in this type of pulmonary problems? The analysis of the ethical aspects of the problem frequently makes evident the need to involve either the patient or the family in the decision process.

After the doubts have been dealt with, the ICU team reexamines the possible actions to be taken in order to check if they still appear as problematic as before, and/or if alternative actions should be considered. Many problems are solved at this point. However, if uncertainties and disagreements persist, each member of the team is asked to state without giving further justifications his personal preferences regarding each of the possible actions. When all the members of the ICU team have stated their options, a set of actions in which all agree upon is identified. It is likely that additional actions also will be suggested in which the team will be only in partial agreement or even total disagreement. The final decision concerning alternative actions where disagreement persists is left to the conscience of the attending physician, to the head of the ICU or whoever is in charge of the ICU when decisions must be taken. Finally, the implementation of the decision is evaluated and discussed.

The principal philosophical judgments or assumptions underlying this decision-making process are the following: 1) the conviction that human actions can be rationally examined; 2) the recognition – in each medical action – of the existence of a technical and an ethical dimension; 3) the objective subordination of the technical dimension of our actions to their ethical fulfillment; 4) the urgent need to guide our actions with the best scientific, technical, anthropological and ethical knowledge available; 5) the recognition of a radical insufficiency of general rational scientific, technical, anthropological and ethical knowledge to adequately apprehend concrete, particular, irrepetible realities; 6) the possibility to attribute different rights and duties to different actors in each situation; 7) the indispensable role of prudence in concrete decision-making about ethical matters; and 8) the possibility to define in some manner an evaluation for this proposed decision-making procedure.

3. A Good Medical Decision

Every effort to improve decision-making in medicine, as well as in any other field, contains the following implicit assumptions: 1) that the decision-making process can be improved; 2) that the improvement can be objectively judged; and 3) that we already possess certain elements on which to base this judgement. If those elements are lacking any proposed alternative to improve the decision-making processes is meaningless. What are then the criteria by which we can ascertain the adequacy of a decision-making process?

A. The Technical Aspect

From a technical point of view, a good decision is that involving the best mean to attain a certain goal. However, 'best' is a relative concept, meaning that the action in question is not a 'good' one in an absolute way, but only the quickest, easiest and most sure way to achieve a desired goal, at a given time and in a given situation (Serani-Merlo, 1989; 1990). This global evaluation can be referred as the 'efficacy' of a given technical action. In general, the technical adequacy of a given action can be judged in accordance to commonly known scientific and technical standards. However, rarely in medicine is a single technical performance involved in a given patient, and rarely are the circumstances in which the technical criteria must be applied, the same as those which were present in the study in which the technical standard was established[10]. This means that the quality of the standards chosen and adequacy of the precise way in which the different technical standards were applied must be ultimately judged by experts. Who are the experts? Which experts must be consulted in a given patient? How many and in which manner? Those questions at present remain unanswered.

B. The Motivational Factor

From an ethical point of view, it is considerably more complex to ascertain what a 'good' decision is. First of all, any theoretical analysis of a 'good' decision-making process assumes that the people involved are well intentioned. That implies that certain common goals are known and shared by everyone (see Israel, 1983, pp.155–193). In real life this may not be the case. Even assuming that only well intentioned people are involved in the decision-making process, most likely not everyone will be attached to the common goals in the same degree. This affective or motivational factor is rarely, if ever, considered in 'rational' decision-making models. In this order of things a good decision is the one reflecting the greatest motivational attachment to the good of the patient from the part of the health care team. By 'motivational attachement' we refer to that frame of mind which most people refer to as 'love'; love to what is being done (philo-tekhné) and love to the people – in this case patients – for whom something is being done (philanthropia).

Can we objectively assess that motivational attachment? It is not an easy task, but there are some ways to approach it. First, we can rely on our own personal judgement. This is a good standard when the sensible aspect of our actions is ambiguous or even equivocal to outside observers. We know, however, that sometimes the nature and the intensity of our motivations is obscure even to ourselves. Also we can rely on the judgment of the patients and their relatives. This can be a good criteria, even if every physician has the feeling – from the time of Sculapius letter to his son – that this cannot be an infallible standard. Finally, we can use the judgment of our peers and the members of the health care team. This has the advantage over the patient's opinion of being unbiased and technically competent. However the limitations of this approach lies in the fact that the observers cannot isolate themselves from their own feelings and motivations. In fact only an observer capable of reading the inner motivations of human beings, thus able to see and understand the real nature and the motivations of our actions could adequately judge about this.

C. General Medico-Moral Goals and Principles

Common goals to be pursued in medical actions are usually referred to – in a deontologically influenced conceptual framework – as 'principles' or – in a natural law conceptual framework – as 'goods'[11]. Principles of medical ethics used to be universally acknowledged, but we all know that this is no longer the case[12]. In our experience, the ethical problems of intensive care are not explicitely refered to questions of 'goals' or 'principles', except for the so-called 'quality of life' discussions. Even when different opinions in questions of general moral principles explicitely exists, clinical ethicists consistently observe that those disagreements do not influence – at least in an obvious way – decisions about very particular and concrete clinical situations. This is also our experience in intensive care ethical analysis. The above does not indicate that moral principles are irrelevant when it comes to deciding on a concrete action. On the contrary, it is only a proof of the irrelevancy of declared disagreements in principles when it comes to the rational discussions of concrete moral facts. Several hypothesis can be proposed to explain this fact. We suggest two: 1) Many physicians act in ways that are inconsistent with their alleged moral principles[13]; 2) A rational discussion can minimize or even suppress many disagreements on moral issues. In keeping with both hypothesis, the observed agreement on concrete options usually reflects a consensus on the most general and most important ethical goal or principle. This consensus is embodied in statements with a universal appeal such as: "Our actions must be in agreement with reason" or "Goodness shown by our rational judgment must be pursued, evil must be avoided"[14]. The above statements cannot obscure the fact that physicians may have real differences concerning their views about the goals to be pursued in medical care. Nonetheless, it is evident that sharing common views is not a *sine qua non* condition in order to reach an agreement regarding the best course of action for an ICU patient.

D. Art and Prudence

The final, and most definitory, factor in decision-making is consumate experience. In the technical or 'artistic'[15] domain this is refered to as expertise, and in the ethical domain as prudence. The opinions and options suggested by both experts and prudents are necessary for an in depth analysis and discussion leading to 'good' decisions. In the decision-making model applied in our teaching hospital, the presence of experts and prudents plays a key role stimulating the discussions, providing additional elements of judgment and, in general, helping the ICU staff to dispell their doubts and finding the right option. The need for experts and prudents is the practical expression of the limitations of human conceptual knowledge. Being compelled to operate under a different régime, when facing a unique constellation of factors, human intelligence is deprived of the help of general laws and conceptual generalities. This is the place for memory, brightness, docility and all the other perceptual, intellectual and affective capabilities that were considered so important by classic philosophers in relation to artistic and prudential performances[16].

Similarly to good technical choices, good ethical choices suppose a relativistic element (i.e., the good moral choice is that which the most prudent man takes in a specific place and time, with the best knowledge available after a conscientious deliberation and in relation to his own responsibility). Absolutely right moral choices, with independence of concrete circumstances do not exist. As Kainoti (1986) has correctly pointed out: "The goal of all moral choices in critical care medicine is to arrive at a good moral choice...rather than the right moral choice". In the moral domain, prudents play the role that experts have in the technical arena, thus prudents are called to advice about the moral 'standards' to be respected in a given situation.

A good medical decision, then, can be tentatively defined as one that, in the judgment of experts, assumes the best scientific and technical standards, and applies them in the most competent way. Such a decision is taken under the guidance of the most rational and noblest goals, by those who are sincerely and intensely motivated by them, as judged by a combination of factors originating in the judgment of themselves, their patients and their peers. Finally, that decision must combine in a synthetic 'architectural' fashion, the technical expertise, the personal devotion for the good, the wide knowledge of moral principles and the prudent application of them to a single, unique and irrepetible set of circumstances. Based on the knowledge of the nature and the limitations of evaluative judgements, it must be acknowledged that an exhaustive, definitive, and absolute evaluation is for us humans only a tentative task.

IV. CONCLUSIONS

Intensive health care is an out growth of modern medicine. The ICU teams take care of patients with either actual or potential high risk of death. This reality implies the need to take technical and ethical decisions that in some cases are extremely complex ones. Presently few concrete and universally agreed upon guidelines exist to orient ICU teams in their decision-making processes and, in this respect, it is important to point out that communities as such have not been involved in the establishement of ICU decision-making policies and guidelines. However imperfect these guidelines could be, they are preferable to not having at all. Improving the knowledge and proficiency on clinical ethics of ICU members may help to better decisions in terms of admissions and discharge of patients from ICU. This type of expertise may also help to improve the quality of patient-physician relationship and would facilitate a more mature and active participation of patients and their relatives in medical decisions.

Decision making in difficult cases can be greatly improved by designing specific procedures for a collegiate deliberation process. They must be conceived not as mechanisms to dilute responsibility but on the contrary to help each actor in the health care scenario to better assume his/her own responsibility. These decision-making models or deliberative procedures must be designed under the guide of the most permanent and realistic philosophical basis, with explicit statement of their assumptions and their strengths and limits.

NOTES

1 The french oncologist, Lucien Israel (1983), has sustained a similar thesis in an extensive essay, applied in his work in a more general context to all those situations in which medical doctors have to cope with decisions in conditions of uncertainty.

2 We use the words 'regulation' and 'control' as they are employed in the theory of control and regulation applied to physiological systems. Cf. Günther & Hodgson (1970) and Günther & Diaz (1980). In our application, however, a physiologically regulated variable could become a medically controlled variable. The inclusion of a medical or

extrinsic controller in a physiological model of regulation and control obligues to precise the exact meaning in which the words regulation and control are used.

[3] This political will in relation to health affairs can be executed at different levels (ministry, region, hospital)and by single persons or in a collegiate way. It is not excluded that a medical doctor could participate in this decision making process. However, that participation is not as a medical doctor but as a medically well informed politician.

[4] This is true from the point of view of social justice. It is clear that other kind of ethical obligations cannot be completely excluded in the sense that physicians and nurses as citizens can, and in certain cases must, participate in political activities, considering the word 'politics'in its widest and noble meaning.

[5] The last two decades have seen interesting and valuable efforts in order to reduce uncertainty in difficult medical decisions. Some of these efforts consist mainly of applications to clinical medicine of decision analysis theories developed in other fields of human knowledge and practice (for a review see Pauker & Kassirer, 1987), while others are models specifically derived from the observation of expert clinicians (Barrows, 1972; Kassirer & Gorry, 1978). It seems that the real impact of these developments is reduced (Detsky, 1987), due both to intrinsic limitations of the models and to sociological inertia. The importance of the utilization of these 'rational' decision-making procedures in medical teaching and in medical practice is still to be defined, but the general feeling of many authors is that they are nowadays underused (Israel, 1983; Kassirer, 1983; Detsky, 1987)

[6] In our times those who know best and actually utilize regularly those criteria are the physicians who work in ICU's and not the attending physicians. This means that the request for a vacant in the ICU supposes – when it is possible – a dialogue between the attendant physician and the ICU team. The latter are the most qualified to know if that kind of patient can be benefitiated on medical terms from intensive care, in what amount and with which probability.

[7] Although the general concepts presented here are applied by us in different clinical settings, the specific process described has been mostly applied in the neonatal intensive care unit of our universitary clinic.

[8] The above statement is an application of classical philosophical ideas to the medical domain (See for example: Von Hildebrand,D., 1983, pp.397–402)

[9] We have developed these concepts in detail elsewhere: (Paz-Henríquez & Serani-Merlo, 1992, pp. 1–13)

[10] Authors in the field of technology assessment refer to this as the different existing between the 'efficiency' of a given technique in the controlled study, and the 'efficacy' of the technique being employed in less controlled but more realistic situations. The studies published by The Rand Corporation of the United States in this field are widely known.

[11] The ethical theory of the four principles: beneficence, non-maleficence, justice and autonomy, derived from David Ross, has been popularized in medical ethics by the group of the Kennedy Institute at Georgetown University (Beauchamp & Childress, 1989). We have traced and criticized elsewhere the philosophical origin of this doctrine (Serani-Merlo, 1992).

[12] The most clear example of this is the open criticism to which the Hippocratic Oath is being subjected in some industrialized countries. Cf. Goic (1992).

[13] This could find in turn its explanation in the fact that most physicians share their

medico-moral principles – in the moment they *publicly* face difficult ethico-clinical decisions – more than what their are willing to acknowledge.

[14] We obviously prefer the second enonciation, because it makes more evident the fallacy of the pretended 'naturalistic fallacy' of Hume. The first moral goal or principle, as it is has been classically stated "good is to be pursued, evil to be avoided" is a judgment of *practical reason* and so it is not an indue intrusion of theory into practical judgments. Moreover, if it is certain that this practical judgment supposes some theoretical knowledge, however, this knowledge is about the 'dynamics' and not about the 'static' of the human being, as Hume seems to assume. We think that the confusion introduced by the 'logical' objection of Hume to natural law theory could have been easily avoided by a more accurate regard to the originality and consistency of moral realities. Agreeing then with Ayala (1987) in his critic to sociobiologists, we do not share his uncritic acceptance of the 'naturalistic fallacy'.

[15] We employ the word art, in this context, as a strict equivalent of technique, as is used to be in ancient greek, arabic and latin philosophers.

[16] A clear-sighted synthesis on this topic, including a critical examination of the opinions of previous philosophers is proposed by Thomas Aquinas (1962) in his *Summa Theologiae*.

REFERENCES

Ayala, F.: 1987, 'The biological roots of morality', *Biology & Philosophy* 2, 235–252.

Barrows, H.S., Bennett, K.: 1972, 'The diagnostic (problem solving) skill of the neurologist', *Archives of Neurology* 26, 273–277.

Beauchamp, T.L., Childress, J.F.: 1989 *Principles of Biomedical Ethics*, Oxford University Press, New York.

Detsky, A.S.: 1987, 'Decision analysis: what's the prognosis'?, *Annals of Internal Medicine* 106, 321–322.

Goic, A.: 1992, 'El juramento hipocrático. ¿Una veneración ciega'? *Vida Médica* 44(5), 25–28.

Günther, B., Hodgson, G.: 1970, *Fisiología Integrativa: introducción a la teoría de control y de compartimiento*, Ediciones de la Universidad de Chile, Santiago de Chile.

Günther, B., Diaz, G.: 1980, *Fisiología de Sistemas 1: Mecanismos de regulación*, Ediciones de la Universidad de Chile, Santiago de Chile.

Israel, L.: 1983, *La decisión médica* (translated from french), Emecé Editores, Buenos Aires.

Kainoti, G.A.: 1986, 'Ethics and Medical-Ethical Decisions', *Critical Care Clinics* 2 (1), 3–12.

Kassirer, J.P.: 1983, 'Teaching clinical medicine by iterative hypothesis testing: let's preach what we practice', *The New England Journal of Medicine* 309, 921–923.

Kassirer, J.P., Gorry, G.A.: 1978, 'Clinical problem solving: a behavioral analysis', *Annals of Internal Medicine* 89, 245–255.

Kilner, J.F.: 1990, *Who lives? Who dies?*, Yale University Press, New Haven & London.

Knaus, et al.: 1991, 'The APACHE III prognostic system: risk prediction of hospital mortality for critically ill hospitalized adults', *Chest* 100, 1619–1635.

Pauker, S.G., Kassirer, J.P.: 1987, 'Decision Analysis', *The New England Journal of Medicine* 316, 250–258.

Serani-Merlo, A.: 1989, 'La lógica de la técnica y la lógica de la ética en la procreación humana' in Caffarra, C. et al. (eds) *Humanae Vitae, venti anni dopo*' Atti del II Congresso Internazionale di Teologia Morale, Ed. Aries, Milano, pp 673–675.

Serani-Merlo, A.: 1990 'El médico frente al impacto cultural producido por las técnicas actuales en reproducción humana' in Lavados-Montes, M. et al. '*Problemas contemporáneos en Bioética*' Ed. Universidad Católica, Santiago de Chile, pp. 209–224.

Serani-Merlo, A., Paz-Henríquez, R.: 1992, 'Problemas éticos en la práctica médica' in Serani-Merlo, A., Lavados-Montes, M. (eds.) *Nociones fundamentales y aplicadas de Etica Clínica* (Syllabus) Unidad de Bioética, Facultad de Medicina, Pontificia Universidad Católica de Chile.

Serani-Merlo, A.: 1993. 'Teorías éticas relevantes en Ética Clínica' in Lavados-Montes, M. & Serani-Merlo, A. '*Etica Clínica: fundamentos y aplicaciones*' Ed. Universidad Católica, Santiago de Chile, pp. 45–64.

Shewmon, D.A., Di Giorgio, C.M.: 1989, 'Early prognosis in anoxic coma: reliability and rationale', *Neurologic Clinics* 7, 823–843.

Strauss, M.J., et al.: 1986, 'Rationing of intensive care unit services: an everyday ocurrence', *J.A.M.A.* 255, 1143–1146.

Thomas Aquinas: 1962, 'De Prudentia' *Summa Theologiae*, Quaestio 47, 48, 49, 50 et 51 (textum editionis criticae Leoninae), Paulinae, Roma.

Von Hildebrand, D.: 1983, 'Etica' (translated from german), Ed. Encuentro, Madrid.

HERMAN H. DELOOZ

ETHICAL ISSUES IN CRITICAL CARE: CRITERIA FOR TREATMENT

I. DEFINITION AND CHARACTERISTICS OF CRITICAL CARE FACILITIES: HISTORICAL BACKGROUND

The first long-term application of life-sustaining therapy in the form of intermittent positive pressure ventilation was accomplished during the last polio epidemic in the Scandinavian countries in the early 1950's. This epidemic lead to the development of the Engström respirator and the acceptance of intermittent positive pressure ventilation as a safe and effective way of sustaining life in patients, whose spontaneous breathing capacity was lost due to paralysis of the respiratory muscles. It was considered essential to bring these patients together in a dedicated area of the hospital and to assure continuous surveillance, although this was, for the majority of time, provided by lay volunteers. A spatial and functional structure providing life-sustaining therapy was thus created.

Soon the experience with long-term controlled ventilation would be applied on non-polio patients still maintaining breathing capacity but considered to be in respiratory failure, secondary to severe chest injuries, others requiring therapy which included abolishment of spontaneous ventilation, such as patients suffering from severe tetanus.

Out of a completely different angle came the proposal by Halderman in 1959 to group patients according to the degree of nursing care required into "high care", "medium care", and "low care" patients. In the view of Halderman, the patients would pass through a system of "progressive nursing care". This strategy was intended to meet the shortage of nurses, based on the hypothesis that grouping the patients would save on nursing capacity. Patients whose condition required intensive nursing care, were brought together in a dedicated spatial and functional structure; life-sustaining medical therapy through the application of special technical devices, such as ventilators, were not necessarily part of this categorisation.

Finally the most important impulse for the development of intensive care was given by the development of open-heart surgery in the 1950–60

K. Wm. Wildes (ed.), Critical Choices and Critical Care, 79–101.

period. Cardiac surgery followed the introduction by Cournand in 1942 of cardiac catheterization, which permitted precise diagnosis of heart conditions, and the rapid development of anaesthesiology as a medical discipline in the Anglo-Saxon countries during the last World War. If however these developments permitted the successful performance of surgery on both congenital and acquired cardiac ailments, it was highly frustrating to witness the death of many patients in the post-operative period. Straight forward surgical logic gave birth to the idea that since the patient was able to sustain the aggressive of the surgical procedure under the care of the anaesthesiologist, there should be no reason why a patient would die, once the "curing" surgery was performed, providing the same care would be continued during the post-operative period. Anaesthesiologists agreed that their way of caring for the patient during surgery might be of benefit after surgery, but this required the spatial and functional facilities to organize and provide such care. This meant a unit with similar equipment to that used in the operating theatre, a dedicated nursing team and an increasing number of anaesthesiologists. They applied a way of practising medicine which up to that moment had been exclusively restricted to the operating theatre: the continuous bedside adaptation of therapy to the changing condition of the critical patient in a "titrated" way.

Thus the characteristics of intensive care can be summarized as care provided in a spatial and functional structure, using all means to evaluate (monitor), to support and, when necessary, to sustain the vital functions of the patient and this through the "titrated" application of all possible technical and pharmaceutical means.

Through its historical background the development of admission criteria to intensive care followed very divergent patterns. When we consider the cardiac surgery patient, we witness the admission to be justified either by the critical condition of the patient at the end of surgery, by the assumption that continued "titrated" care will improve outcome and diminish morbidity, or by the wish to control the risk of complications. This last way of justifying admission to intensive care is in most hospitals not applicable to patients after other types of surgery, such as abdominal surgery, although experience seems to indicate that "preventive" admission of these patients can reduce mortality and morbidity. When risk is appreciated, two factors are taken into consideration: the frequency of the complication and its gravity. As such the objective risk equals frequency times gravity. Subjective appreciation of risk however will attribute a certain power to the factor gravity. Subjective risk then equals frequency times gravity to the power. It is the power given to the gravity of the complication, that supports the decision to admit the post cardiac surgery patient rather than the post abdominal surgery patient. The heart being "the" vital organ gives extra power to the factor gravity. However this attitude is not supported by any controlled data. Most probably the role of cardiac surgery played in the coming about of intensive care, also offers a "historical" argument in favour of the cardiac patient's admission to intensive care.

Soon after the introduction of intensive care in the post cardiac surgery phase, cardiologists found out that mortality after myocardial infarction was largely due to arrhytmias. Consequently they decided that close observation of these patients and their cardiac rhythm, followed by immediate intervention when life-threatening arrhythmia occurred, could cut on mortality. The coronary care units were considered as another kind of intensive care. However most of the patients admitted to these units, only need intensive observation and monitoring, but neither intensive nursing, nor intensive titrated medical treatment.

Neonatal intensive care concerns mainly premature underweight babies. The ethical issues involved herewith are even more complex than those involved in adult intensive care, for the uncertainty of outcome as far as quality of life is concerned, reflects on a total life span, on a handicap weighing on the individual from the start of his life, as well as on his relatives and environment.

Different forms of intensive care not only have introduced different admission policies, but also different organization schemes. The true intensive therapy units almost always have their own dedicated medical staff, which is fully in charge of the patient, although the referring physician or surgeon is supposed to give advice in matters concerning his/her specific competence or to the extent of care to be given to the patient. In this kind of unit the head of the department together with his staff will elaborate and implement admission and discharge criteria. Their continuous involvement with the care of critical patients makes them the true experts in the diagnosis, therapy and prognosis of the critically ill or injured. Units that mainly offer observation or intensive nursing care on the contrary, may allow any physician from the hospital to be in charge of the patients, which more or less excludes regulations as far as admission or discharge are concerned. Even if some rules are made up by a hospital committee, the application will not be controlled and will mainly be dependent on the political power of the physician in charge of the patient.

We consider any type of intensive care important enough, as far as costs are concerned, to deserve a medical authority in charge, not only because of the need for admission and discharge criteria, but even more because of the necessity of specific quality control and audit. For if intensive care provides the patient with extra chances of survival and recovery, it also threatens the patient with complications due to the invasive nature of the techniques applied and the power of the drugs and antibiotics used.

A more recent evolution has brought the critical care approach to the entrance of the hospital: the Emergency Department. The overall goal of Emergency Medicine is to create a spatial and functional structure dedicated to the care of all medical emergencies. Top priority within this structure concerns the limitation of mortality (both early and late), morbidity (both in gravity and duration) and invalidity, through the **early** diagnosis and therapy of all life-, organ- and limb-threatening conditions, caused by illness or injury (Delooz, 1992, p. 1–23). In this perspective time is of extreme importance.

The evolution of Emergency Medicine has only taken this form over the last ten years, but will certainly become general, as oxygen debt (the cumulative oxygen deficit over time) is increasingly recognised as the main factor determining organ failure and thus morbidity and mortality. Only the continuous availability of expertize in critical care through the development within the department of an observation care unit including facilities for short term intensive care, will assure the presence of the necessary expertize at all times (Delooz, 1983, p. 129–130). As life sustaining care will often be applied even before a causal diagnosis is reached, decisions to stop resuscitation or intensive care will have to be made as well as decisions to forego critical care. Therapeutic trials can thus be carried out under close supervision by the same team which has admitted the patient. Here also a specific medical staff trained in emergency critical care becomes indispensable.

From this overview of the development of critical care (intensive care and emergency care), we conclude that this type of medical care should not be organized on an open basis, but deserves a dedicated medical team and authority, to assure the quality of care that should be expected, taking into account the economic means involved and the ethical aspects of its use.

II. FOREGOING THERAPY – 'ACHARNEMENT THERAPEUTIQUE'

The French term "acharnement thérapeutique" can be defined as followed: a medical attitude characterized by the application of therapeutic means which are not in proportion to the gravity of the illness or injury, or the expected result. As such, this definition includes the application of diagnostic and therapeutic procedures which are out of proportion to the rather benign or trivial nature of the illness or injury.

It also includes the application of therapeutic means which are not in proportion to the result that can be obtained as far as adding to the life of the patient in terms of both quantity and quality.

The second part of the definition is the subject of this section of the paper.

Although each case is unique, but also because each case is unique, we want to tackle this subject by giving two short case reports and discuss the bases of and the modalities for decision making in each of them.

Case 1: a man of 68 yr of age is suffering from chronic respiratory insufficiency to such a degree that he has spent the last six months as a complete invalid between his bed and his chair. A sudden attack of acute respiratory insufficiency brings him in coma to the Emergency Department. There is no time for debate, immediate endotracheal intubation and controlled ventilation are required. The sudden deterioration of his respiratory status appears to be due to a pneumonic infiltrate of the right upper lobe. After adequate antibiotic therapy, the upper lobe consolidation persists and a trial at weaning from the respirator is unsuccessful. The patient who regained consciousness soon after the start of artificial ventilation, communicates in writing and, on his request, is informed that a bronchoscopy is planned. It might be either of therapeutic value, if it permits opening up the bronchus to the right upper lobe, or it might give diagnostic information. The patient's respiratory status however makes a bronchoscopy a rather hazardous intervention, since it will interfere with optimal controlled ventilation. The patient expresses the wish to go ahead even considering the risk. Bronchoscopy reveals a malignant tumour at the entrance of the right upper lobe. As requested by the patient, he is informed of the result of the bronchoscopy: the right upper lobe cannot be restored to its function. The patient concludes that in this case successful weaning from the respirator would be highly implausible.

When the patient's conclusion is asserted by the medical team, he asks to stop controlled ventilation and to let him die in peace.

After information and consultation with his family (one of his sons is a physician) and having allowed ample time for reconsideration, the patient's request is fulfilled. He receives enough oxygen and a minor amount of morphine to avoid dyspnea and to ensure his comfort while dying.

This case is characterized in the first place by the fact that often in emergency situations supportive intensive therapy has to be started before diagnostic procedures are carried out. This implies that reflection on the ethical implications of medical activities within this context is mandatory. Furthermore, the diagnosis and the immediate prognosis are clear due to the previous chronic condition of the patient and the nature of the acute deterioration. The patient could be informed and was competent to participate in the decision making process. Finally stopping controlled ventilation meant that the process of dying was shortened, while continuing controlled ventilation would have added quantity of life, but according to the patient's own interpretation, no valuable quality of life. Although controlled ventilation is stopped, the patient continues to receive medical attention and care.

Case 2: The second case concerns a young man of 20 who at 8.30 am is brought by his parents to the Emergency Department in a terminal state of respiratory insufficiency. His arterial oxygen tension is below 30 mm Hg, his arterial carbon dioxide tension is above 90 mm Hg. The patient is stuporous and shows gross signs of right heart failure.

The boy is the third child of a family of 5 children; the first two are healthy, the third and the fourth are suffering from cystic fibrosis. The last hospital admission of the patient goes back to the age of 12. At that time the diagnosis of cystic fibrosis was made and the boy decided that he would not go back to the hospital and preferred to be taken care of by the family physician.

That morning however, he was brought to the Emergency Department on his own request. The condition of the patient was so poor that controlled ventilation was judged to be indicated by the residents who first saw the patient. Since however this patient was clearly in a state of irreversible respiratory insufficiency without any acute cause of deterioration, we considered the patient not to be a candidate for controlled ventilation, nor intensive care. In view of the poor physical development of the young man and the broad experience with the disease in Pediatric practice, consult from Pediatric staff was asked, in order to discuss further diagnostic and therapeutic strategy. A younger staff member, and later the chairman of the department of Paediatrics, advised against controlled ventilation and in favour of the necessary care to provide a peaceful dying process. The residents however argued that the patient might not have had optimal therapy since his last admission and requested permission to at least optimise conservative therapy. We agreed that there was no objection to give a therapeutic trial with titrated oxygen therapy, respiratory physiotherapy, diuretics and cardiotonic drugs. At 8 pm, this therapeutic trial was without any effect and the discussion on the use of controlled ventilation was started over by the residents. After consultation with the chairman of Paediatrics, we concluded that controlled ventilation might have a positive effect on the right heart failure and this within a period of 24 hrs. We discussed this possibility with the parents of the patient and agreed that controlled ventilation would be applied for 24 hrs. After that period the right heart failure would either have responded, which would mean that the maximum benefit had been achieved, or it would not have responded, which would make continuation of controlled ventilation physiologically futile.

In either case artificial ventilation would be stopped after this therapeutic trial. The parents agreed with this proposition and controlled ventilation was started that night and continued for about 36 hrs, to permit weaning during day time. During the period of controlled ventilation the patient lost 9 litres of fluid and regained perfect consciousness, allowing him to see his family and an elder neighbour with whom the boy spent a lot of time.

After stopping the controlled ventilation, the patient returned very fast to a state of stupor, comparable with his condition on admission. His parents however valued highly the contact they had with their son during the period of controlled ventilation.

The next morning the young man was not only alive, but his mental state had improved in a way that he was able to make the well known sign with his thumb in the air, when I was inquiring about his condition. He was transferred to an ordinary ward, was discharged after a week and obtained his driving licence and actually drove his car two months after discharge. He died at home about two years later.

This case is characterized by the fact that ample consultation was possible before intensive therapeutic action had to be taken. It illustrates the relative value of a "generally accepted therapeutic attitude", for all textbooks in those days were referring to controlled ventilation in the terminal state of cystic fibrosis as physiologically futile. It also illustrates that although prognosis may be well established, it may not hold true for the immediate future in cases of chronic illness. Quality of life is completely personal and can only be assessed by the patient himself, except for the patient in persistent vegetative state. A therapeutic trial should be limited in time and the relatives should be informed as well on the goal of the trial as on the time frame. The family has to be fully informed, but the decision to offer a certain therapeutic alternative is made by the physicians in charge as a team and after ample consultation.

Dying, even in its most acute form, (e.g. sudden cardiac arrest) is a process. Bringing this process to a halt and, if possible, reversing it, is the ultimate goal of critical care. This means that eventually critical care could be applied to every person, whereas other forms of medical therapy are only applicable to patients suffering from a specific disease or a specific injury. This consideration makes clear that even if critical care would be available to all at all times, it should not always be applied. Critical care has to be foregone whenever it is not beneficial to the patient in terms of adding both quantity and quality to the life of the patient. The basis for this decision is mainly the prognosis of the case and the expected effect of the therapeutic intervention.

Prognostic scores, such as the APACHE scores have been developed, but as stated by W.A. Knaus: "Currently available, numerical mortality estimates must be supplemented with mostly subjective estimates of quality of life. The characteristics and the continuous nature of the estimates must be emphasised, lest objective estimates be misunderstood as decision rules, which might restrict rather than enhance clinical reasoning. Objective probability estimates will not resolve most ethical controversies"(1991a, p. 389–394).

Only "in a few cases, risk estimates could support clinical judgement that continuing current therapeutic efforts would be futile" (Knaus, 1991b, p. 1619–1636). Or as stated by R.D. Truog *et al.*: "Even if we could agree on a statistical cut off point for determining futility, physicians are often highly unreliable in estimating the likelihood of success of a therapeutic intervention"(1992, p. 1560–1564).

In an effort to define futility during cardiopulmonary resuscitation, we studied an early prediction score (EPS) as the sum of five events that are clearly correlated with a favourable outcome: the type of cardiac arrest is ventricular fibrillation; the type of respiratory arrest is gasping; pupil

		Diagnosis: clinical yes	no
Diagnosis: biopsy necropsy	yes	a	b
	no	c	d

Fig. 1.

reaction is unequal, slow or normal, but present; swallowing or gagging activity is present and the cardiac arrest has been witnessed. Presence of any of these events contributed one point to the score, while absence contributed nothing to it. EPS studied on 2235 consecutive cases of cardiac arrest, during resuscitation, resulted in a comparable amount of information, whether used to predict success (alive and conscious 14 days post CPR) or no-success. However EPS during CPR did not allow any decision making as far as stopping or continuing CPR efforts, for an EPS of zero yields a probability for no-success of 0.96 (Delooz, 1989a, p. S149–S155). If however the probability of no-success would have been 1.00, we would still have considered this study to be of limited value as far as decision making is concerned, for it would only apply for the patients in our study group and cared for by our Emergency Medical Service System. And, what is even more important, this study, as all outcome studies, has to be relativated, because of the aspect of *self-fulfilling prophesy*, which is inherent to all outcome studies.

As a matter of fact, whenever a comparison is made between a clinical diagnosis and a diagnosis reached by biopsy or necropsy, true positives and true negatives can be clearly differentiated from false positives and false negatives (see Figure 1, the number of false positives is represented by c; the number of false negatives represented by b).

However when a study of prognosis as compared to outcome is carried out, true negatives (a in Figure 2) and true positives (d in Figure 2) can not be considered as such, because there is always contamination between the prognosis and the final outcome and this in both directions; therapeutic efforts will consciously or unconsciously be modulated by

		Prognosis: alive	death
Outcome:	alive	a	b
	death	c	d

Fig. 2.

the appreciation of the chances of survival made by the team caring for the patient.

This means that objective prognosis can not serve as a criterion for decision making to forego certain therapeutic acts and/or critical care. However it does not mean that we consider such decisions as impossible to be reached. The examples given, illustrate that we believe, they should be made.

Physicians may acknowledge that therapy is effective, in a limited sense, but believe that the goals that can be achieved are not desirable, as when considering prolonged nutritional support for patients in persistent vegetative state (Schneiderman, 1990, p. 949–954). The decision that certain goals are not worth pursuing is best seen as involving a conflict of *values* rather than a question of futility (Truog, 1992, p 1560–1564).

As far as the role of the physician is concerned, we do not believe in the "paternalistic" role, which implies that the doctor judges and makes the decisions without informing or discussing the future with his patient. A physician-patient discussion and shared decision making is to engender a sense of trust between patient and physician (Walker, 1991, p 2407–2433). We neither believe in total "autonomy" of the patient as the model of choice. The USA medico-legal system has favoured this model, but more as a means to avoid legal responsibility, than as a result of ethical reasoning.

Some physicians allow patients (or surrogates) to decide when a treatment is futile, thereby overriding medical judgement and potentially allowing the patient or surrogate to demand treatment that offers no benefit (Schneiderman, 1990, p 949–954). The physician should benefit the patient according to the most objective judgement available, unless the patient expresses a competent and informed wish for an alternative course (King's Fund Panel, 1989, p. 428–431).

In cases in which competence cannot be assessed, decisions must rest with the doctor, but always in the context of close consultation with the family and a presumption in favour of the preservation of life must predominate. Conversely, there is no moral or legal obligation to provide treatment on request, when there would appear to be no possibility of benefit (King's Fund Panel, 1989, p. 428–431).

In many cases, there is a critical point beyond which medical interventions may act less to prolong acceptable life, than to extend a miserable dying process. Professionals cannot expect patients or families to take the lead in raising these questions (Ruark, 1988, p. 25–30).

We consider the model of "mutual respect" between the patient and his family on the one side and the medical team on the other, to be the model for information and decision making, rather than the two extremes "paternalism" and "autonomy" and would therefore endorse the statement made by the Medical Section of the American Lung Association (1991, p. 726–731): "It is important that the health care team continues to participate actively in making health care decisions on the patient's behalf in conjunction with the patient's surrogate" or as it has been put by R.M. Walker (1991, p. 2407–2433): "... An opportunity for physicians and patients to discuss their views ... preserves physician's autonomy and allows patients to exercise theirs".

Coding Decisions

Too often DNR coding is used to indicate a therapeutic attitude much more conservative than the "do not resuscitate" code would imply. As a matter of fact, in a recent survey on 25 patients by G.C. Webster *et al.* it was stated that "the DNR order was not accompanied by withdrawal of other therapy in 50 % of the cases and one patient recovered and was discharged from hospital". Nevertheless, the authors add "[The] extent to which physicians and nurses equated DNR order with limiting treatment was unexpected" (Webster, 1991, p. 553–563).

Of course this study revealed that in 50 % of the cases the DNR code implied more than just the indications that no chest compressions and no electrical defibrillation would be carried out in case of a cardiac arrest, which certainly is the original significance of this code.

We therefore have instituted in our department a coding indicating the decisions to limit therapy, which differentiates between the conservative DNR code on one hand, the decision not to broaden the therapeutic approach in general or in a limited fashion, and finally the decision to stop therapy in a general or in a limited specific way.

It is stressed that decisions on the limitation of therapy can only be taken after broad consultation at staff level and after informing the close family of the patient. However, any member of the team, be it a doctor-resident or a nurse, can take the initiative to start the discussion on coding. Broad consultation amongst all doctors involved in the care of the patient, including the referring physician, is advised.

The decision is explained and clearly communicated to all involved in the care of the patient and unequivocally posted on the patient's record, in order to assure transfer of information among the different medical and nursing shifts. Of course any decision can be reviewed at any time, but again only at medical staff level.

No code means that the patient will receive all care that is felt to be indicated.

Code I: means DNR, or no thoracic compression, no electrical defibrillation in case of circulatory arrest

Code II: means no extension of therapy in the general sense, which implies no new therapeutic scheme will be started, or in a limited sense, which means that certain, well-defined therapeutic acts will not be added to the therapeutic scheme of the patient.

Code III: means that intensive therapy in general will be stopped or that certain specific therapeutic acts will no longer be part of the therapeutic scheme.

In the medical record of the patient the motivation for the coding and the extent in cases of code II and III must be clearly outlined. This note has to carry date and hour as well as name and signature of the staff member who made the decision after consultation.

This way of coding foregoing of therapy aims at an unequivocal transmission of information between members of the medical-nursing team as well as between the team and the patient and his family.

To conclude, we want to state that the purpose of life-sustaining intervention should be to restore or maintain a patient's well-being and it should not have as its goal the unqualified prolongation of a patient's biological life (American Thoracic Society, 1991, 726–731). If possible the patient should be part of decision making. The family should be informed but not decide for the patient. Physicians as a team will judge on the basis of all the information available. If therapeutic trials are considered, a time-limited trial of the intervention where upon is agreed by all involved, may increase the certainty of the prognosis.

The role of ethicist's does not involve concrete decision making, but concerns the study of the circumstances and issues at stake and the instruction of the physicians as to the criteria and procedures to follow in decision making.

III. THE ETHICAL APPROACH TO COSTS AND BENEFITS IN CRITICAL CARE

Medicine's technological miracles often raise the toughest questions mankind could face. By offering new extensions of mortality, they present circumstances never encountered before. Artificial life-support capability gives a second chance to critically injured persons. It also keeps irreversibly comatose patients "alive" indefinitely. Neonatal intensive care may provide a normal life to three-and four-pound premature babies who would otherwise have died. It also salvages one-pound new-borns with permanent impairments. Inevitably the questions arrive: When does life end? Is life always worth saving? The answers come individually and often quietly, primarily reflecting concern for the patient's suffering, dignity, or expected quality of life. However, as if these questions are not difficult enough, cost issues are adding a new twist (Richards, 1984, p. 80–86).

Both government and private employers are reaching a spending limit for health care. In a closed pool of resources, this means that every time access to the resource pool is granted, it could be denied to someone else, or to quote David Owen: "The nature of health available to one patient can influence the quality and content of care available to another"(1984, p. 1341–1345).

An example is the study carried out in the USA in 1977, which showed that by lowering the symptom severity for undertaking appendectomy enough to save one life, it would be necessary to perform so many operations on normal patients that there would be an added cost of 43 million US dollar and a loss of 2053 person years in convalescence (Owen, 1984, p. 1341–1345).

Faced with the reality that costs equal foregone benefits elsewhere, physicians can take one of two attitudes:

A. They believe that *physicians have a special role-related ethic,* that exempts them from the general ethical duty to promote fair and reasonable distribution of resources (Veatch, 1981, p. 13–14).

The physician-patient relationship is considered to be the moral center of medicine or as Beauchamps and Childdress (1983) have put it: "Physicians are to do all they can for their patients, without taking into account the kind of factors ... that policy makers rightly should consider". If choices have to be made, let society, the social security department, the ministry of public health, the hospital administration tell the physician what not to do. It will be their responsibility. The physician will worry exclusively about the patient and the hospital has to look after the costs. In other words, someone else has to decide.

Of course society has a role in the provision of adequate health care: *The natural and social lottery cannot be denied* (Engelhardt, 1986, p. 1159–1164). This however does not imply that the effects of these lotteries have to be accepted as such. Society has the moral duty to correct the effects of these lotteries to a certain extent, through the assurance of equal access to basic human needs, such as health care, including critical care when required. If this correction is not provided, the negative effects of natural and social lottery will amplify each other and will produce through a vicious circle a downward spiral leaving the subject without any prospect. Society can only achieve this goal through an insurance system based on solidarity, in which all citizens are required to participate according to their revenue. Of course such a system, while assuring access to basic needs, such as education and health care, in order to correct for natural and social lottery, should preserve an acceptable difference in revenue and financial status, in order to safeguard human incentive.

Society, however, can only fulfil its role when it is informed of the real needs to be provided in health care.

The completeness and accuracy of any decision is fundamentally dependent upon the collection, interpretation and communication of information. Wherever gaps occur in this process, ethical uncertainties may appear (Gregory, 1984, p. 102–119). This information will have to be offered by physicians, otherwise they will face an inevitable curtailment of their clinical freedom. Maintaining clinical freedom necessitates the involvement of physicians in the process of choice and in the ordering of priorities (Owen, 1984, p. 1341–1345). We will come back on this issue in the latter part of this section (B.4. collectively revising protocols).

B. The second attitude *physicians* can adopt, is to *play an active role in cost-containment.* This can be done in 4 different ways (Morreim, 1985, p. 30–38):

1. *The hospital restricts the physician.* The hospital can monitor physicians spending patterns over time and, where high-spending physicians refuse to curtail their ordering habits, revoke their privileges to practice at that hospital. It can identify the most expensive products, services and technologies, and place restrictions on their use. Or, where a physician is judged to over-use a particular technology, the hospital might monitor or restrict his privilege to use that particular technology (Veatch, 1981). This, however, may be unfair to physicians and patients. Physicians who treat a higher proportion of elderly or high-risk patients, may well need to utilize a higher proportion of the hospital services. Medical care may suffer. This plan, however undesirable, may deserve limited room: surely we cannot find fault if a hospital identifies particularly expensive forms of care and educates physicians about the financial implications of their decisions (Morreim, 1985, p. 30–38).

2. *Physicians share the profits.* At first sight this seems to be very unethical, although we should not forget that in a traditional fee-for-service system, the emphasis on the almost "sacred" quality of the physician-patient relationship, is not entirely altruistic (Thomasma, 1985, p. 204–209). As a matter of fact in the fee-for-service system, the basic rule will be "the more I do, the more I make". Or, if the physician is on a salary "the more I do, the more the hospital makes". But even in this latter case, attention to the hospital profit rarely is entirely altruistic.

Several systems have emerged recently in an effort to win the physicians for cost-containment efforts by making them sharing the profits. The diagnosis-related-groups-prospective-reimbursement-plan (DRG) provides general policies about reimbursement, coverage and eligibility. Within this frame the hospital determines how much it can afford to offer and to whom, and since physicians control 60 to 80 % of the hospital cost through their medical decisions about the use of the resources – whom to admit, how long the patients will stay, what number and type of ancillary services will be used (Spivey, 1984, p. 984–986) – their participation in the plan is essential. The health maintenance organizations (HMO) will offer all the required medical care to a group of employees, provided they pay a fixed monthly contribution. These organizations have shown to provide care with 40 % fewer hospital admissions and at 25 % of the cost, when compared to a fee-for-service system (Stein, 1985, p. 14–18). Preferred-provider-organizations (PPO) will give a discount on health care in exchange for access to a large group of patients who will exclusively make use of their health care facilities. Again these organizations have shown to be able to offer health care with a price 25 % lower than a regular fee-for-service plan (Stein, 1985, p. 14–18).

Of course the basic rule in all these systems is the opposite from the one that governs the fee-for-service system. It now sounds "the less I do, the more the hospital makes, or the more my company makes, or the more my group-practice makes". Such systems can be very successful as far as cost-containment is concerned, but the success of such efforts may well mean the end of the role of the physician as independent patient advocate. This has negative consequences for all patients, but is potentially disastrous for those in critical care units. The human price tag will be very high, for access, quality of care, medical progress and the autonomy of hospitals will all be sacrificed in the name of *efficiency* (Doulenc, 1985, p. 19–29). As the demand for efficiency accelerates, the patient will more and more come to be regarded as a product. As a matter of fact efficiency means doing things properly to meet the norm. This norm is merely financial and as far as the product in itself has no ethical dimension, efficiency as a rule of conduct can be acceptable. However, in patient care the product has an ethical dimension and meeting the financial norm can never be the only standard.

3. *Bedside budget balancing through selection of patients*. The President's Commission for the Study of Ethical Problems in Medicine and Biomedical and Behavioral Research, has stated "Society has an ethical obligation to ensure equitable access to an adequate level of health care for all without excessive burdens" (Richards, 1984, p. 79–86). At first sight this statement seems to be very rigorous. Nevertheless "equitable access" does not mean "equal access" and the commission does not give a definition of an "adequate level of health care", neither of "excessive burdens". Patient selection could mean that the critically ill, the elderly, the poor are left for another institution, or else that certain types of care can be denied to mentally retarded, to mentally disturbed or to patients above a certain age, for example the age of 70. Whenever patient selection is carried out, there will be a bias involved. For instance if the criterion would be age, putting an age limit to the access to certain types

of care, would involve a bias against the elderly. However, if no age limit is applied, the bias would be against the younger patients, as caring for the elderly might have as a consequence, that certain types of care would more likely be denied to younger persons with a much higher and longer expectancy of life. As Morreim (1985, p. 30–38) puts it, the proper concern for securing benefits at the lower cost, can easily slide into a concern whether the patient's benefit itself is worth the cost to society. The physician is then inquiring not about whether the benefits merit the cost, but whether the patient merits the benefit (Morreim, 1985, p. 30–38) and rightfully the question may be raised as to whether the physician can compare benefit of therapy with remote uses of money.

4. *Collectively revising protocols*. Collectively revising protocols as a means of cost-containment involves re-examing the idea that, if we can do something in health care, we should do it. As a matter of fact, instead of drawing the line at the high-cost items, we should draw the line much earlier, at those items that have no real benefit. Intensive care, more than any other type of medicine, has believed that as soon as technology became available, it had to be included in daily practice. At the same time we have witnessed in critical care medicine an almost autonomic transition of procedures from the experimental to the therapeutic categories. We must now direct our attention toward eliminating wasteful practices, thereby conserving resources without seriously threatening quality (Delooz, 1983, personal communication[1]). Careful audit is needed, particularly when procedures initiated for diagnostic purposes are continued as a means of monitoring. This progression is only justified if there is a significant risk of change and earlier or more accurate recognition of it would influence outcome (King's Fund Panel, 1989, p. 428–431).

Effectiveness is the key to this type of cost-containment. Effectiveness requires doing the proper things to achieve the objective, which involves the quality of care and the maintenance of the physician-patient fiduciary relationship.

If we look at the physician's actual behavior, we see that it is guided considerably by the formal and informal protocols, that inevitably emerge in their daily clinical practice.

No need to say, that more medicine does not necessarily equal more health. In the USA the life expectancy of adults has not improved in the 32 years during which national expenditure on medical care has risen from 12 billion dollar in 1950 to 275 billion in 1982 (Owen, 1984,

p. 1341–1345). Longer hospital stays do not necessarily mean better care, as the chances of nosocomial infection in the hospital are 3 % (Stein, 1985, p. 14–18). Where sepsis involves 1 % of the patients in the hospital, it threatens 7.4 % of the patients in the ICU.

The practice of putting down Swan-Ganz catheters, at a cost of 200 dollars per thermodilution catheter, and at a rate of over 1 million Swan-Ganz catheters per year over the world, involves, according to E.D. Robin (1985, personal communication[2]), a 1 % mortality and a 15 % complication rate. At the same time the usefulness of this procedure in patient decision making is questionable. As an example, Swan-Ganz catheters are accepted practice in the differential diagnosis of pulmonary oedema. But nevertheless all experts are unanimous to state that the only therapy for pulmonary oedema is lowering the left atrial pressure, and when questioning the level of left atrial pressure to attain, the answer will be to lower it as far as can be tolerated. In addition the extra possibility of continuously monitoring the mixed venous oxygen saturation, increases the cost considerably, but has only provided the users with one additional piece of information: beginning sepsis is marked by the disappearance of the usual correlation between cardiac output and mixed venous oxygen saturation.

A study by C.F. Miller and collaborators (1984, p. 318) about the utilization of arterial blood gas determinations in a surgical intensive care unit, showed that only 26 % of the routinely determined blood gases were followed by any therapeutic change, whereas this percentage is increased to 55 % if the arterial blood gases are drawn on demand. The conclusion was taken that routine orders of arterial blood gases at timed intervals are rarely useful in either documenting or predicting problems in the absence of clinical signs.

Do we really have to accept the standing order tests in an intensive care unit to determine the serum potassium level every 4 hours. No theoretical study can be shown, documenting that this timing is in accordance with the evolution of potassium levels in the serum and thus is superior to measuring potassium levels every 6 hours, 12 hours, 24 hours or for that matter every 5 or 30 minutes. Of course no practical evidence is available to prove that routine measurement of potassium levels has any influence on morbidity or mortality.

A survey of the literature on the usefulness of patients monitoring published as an editorial by N.L. Pace (1985, p. 197–200), showed that from 1966 to 1984 not one outcome study available tested, in a controlled clinical trial, the value of monitoring in anaesthesia. This lead to the conclusion that "... new monitors should be adopted if they can plausibly improve the standard of patient care... We must remain sceptical of any new monitors until controlled clinical trials can be designed which will test their actual merits".

Intracranial pressure (ICP) monitoring for patients after heavy cerebral trauma is widely applied, although not one study so far has been able to prove that this practice has had any influence on patient outcome. On the contrary, a comparable patient population cared for in a centre which does not monitor intracranial pressure, shows equal outcome when compared with another centre which routinely monitors this pressure (Jennett, 1990, p. S15–S19). Nevertheless ICP monitoring is costly both in technical equipment and in manpower and carries a definite risk of infection for the patient.

These are only a few examples to illustrate the need and the room for revising protocols. Collectively revising protocols is not only a must because of economic pressures. Eliminating useless practices out of the intensive care environment is an ethical obligation towards our patients and society. Practice has for too long followed the unwritten rule "the more I do, the better I do".

Collectively revising protocols will be economically effective. It will trim our expenditures, while preserving quality of care. When it will occur, genuine scarcity will be recognisable. Since it is a collective enterprise, protocols will be open to inspection and public review. The physician-patient fiduciary relationship will be preserved and quality of care may even improve by reducing both the iatrogenic injuries and diagnostic false-positives (Capron, 1983). Or to quote A.H. Morris: "Standardization of care through protocol application based upon credible outcome results will likely simplify therapy, improve outcome, and enhance the physician-patient encounter. By contributing to credible outcome research results, it will likely permit the identification of ineffective therapy and thus set the stage for the elimination of resource consumptive interventions that do not contribute to favourable patient outcome"(1993).

The medical profession and intensive care in particular, have long been vigorous in researching which new things physicians should be doing for their patients. We now have to research, just as vigorously, what needs *not* to be done (Morreim, 1985, p. 30–38).

Such research should not only aim at the characterization of the condition of the patient and the definition of outcome in terms of survival, but provide a tool for auditing the system and the process of care.

One example is the recently proposed "Utstein style" for reporting on cardiac arrest and cardio-pulmonary resuscitation. Our experience with registration of cardio-pulmonary-cerebral resuscitation had lead to the conclusion that uniform definitions, not only for outcome, but also for time-intervals and expertise offered, were essential in order to allow comparison of different systems and define the properties of each system which were instrumental in determining the outcome (Delooz, 1989b, p. S199–S206).

This conclusion lead to a co-ordinated international effort which has produced the Utstein style of reporting (Chamberlain, 1991, p. 1–26).

A first application of this methodology was made by D. Mauer and collaborators (1992, personal communication[3]) to evaluate within the German Emergency Medical Service (EMS) system, the use of semi-automatic defibrillators, and lead to the following conclusions:

1. "Utstein style" not only means uniform documentation, but also quality assurance: *structure*, *process* and *outcome* can be analyzed with the help of clearly defined time intervals and definitions.

2. Semi-automatic defibrillators improve outcome, provided they are used on subjects of cardiac arrest, who receive basic life support within 4 minutes, who are defibrillated within 8 to 12 minutes and who receive advanced life support within 4 to 12 minutes after defibrillation.

This type of research will avoid that semi-automatic defibrillation will be added to whichever EMS system, which would increase costs without proportionally improving outcome.

Another example are the Major Trauma Outcome Studies (MTOS) as they have been initiated in the United Kingdom by the University of Manchester Department of Accident and Emergency Medicine and in Belgium by the University of Leuven Department of Emergency Medicine. These studies are based on the MTOS initiated in the USA by Howard Champion and co-workers (1990, p. 1356–1365), which characterizes the patient's injury through scoring of its effect on anatomy and physiology and correlating these scores with survival. A central objective of the MTOS was the identification of patients whose unexpected outcomes suggested that their cases would be good candidates for peer review (Champion, 1992, p. 47–66).

The Manchester and Leuven programmes however have broadened the scope of the MTOS: included are the registration of intervention times, level of expertise of care providers both pre-hospital and in-hospital; characterization of the patient's injury and outcome are not only registered in terms of survival but also in terms of morbidity measured through the length of hospital and intensive care admission, total and differential cost of care and finally invalidity at discharge from the hospital or at 3 months after the accident.

These types of research will allow identification of those elements of care which are effective in providing the recovery of the patient and hopefully also of those elements of care which are of no effect and thus futile.

NOTES

[1] President's address to the Belgian Critical Care Society, Gent, February 25, 1983.

[2] Personal communication at the 3rd International Symposium on Applied Circulatory and Respiratory Physiology in Anaesthesiology and Intensive Care, Rotterdam, November 26–28, 1985.

[3] Personal communication during the joint meeting of the European Resuscitation Council Working Group on Research Co-ordination and the European Academy of Anaesthesiology Working Group on CPR, Brugge, October 16–17, 1992.

REFERENCES

American Thoracic Society, Medical Section of the American Lung Association: 1991, 'Withholding and withdrawing life-sustaining therapy', *American Review of Respiratory Disease* 144, 726–731.

Beauchamp, T.L.: 1983, *Principles of Biomedical Ethics*, Oxford University Press, New York.

Capron, A., Executive director of the President's Commission for the Study of Ethical Problems in Medicine and Biomedical and Behavioral Research: 1983, Government Printing Office, Washington DC.

Chamberlain, D., et al.: 1991, 'Recommended guidelines for uniform reporting of data out-of-hospital cardiac arrest: the 'Utstein style'', *Resuscitation* 22, 1–26.

Champion, H., et al.: 1990, 'The major trauma outcome study: establishing national norms for trauma care', *Journal of Trauma* 30, 1356–1365.

Champion H.: 1992, 'Trauma patient scoring', in H. Delooz (ed.), *Emergency Medicine and the Anaesthetist* Vol 6, Baillière Tindall, London-Philadelphia-Sydney-Tokyo-Toronto, p. 47–66.

Delooz, H.H.: 1983, 'The integration of emergency care in a total critical care medicine (CCM) system', *Disaster Medicine* 1, 129–130.

Delooz, H.H., et al.: 1989, 'Early diagnostic indices after cardio-pulmonary resuscitation (CPR)', *Resuscitation* 17 suppl., S149–S155.

Delooz, H.H., et al.: 1989, 'Panel discussion', *Resuscitation* 17 suppl., S199–S206.

Delooz, H.H.: 1992, 'Emergency Medicine: an anaesthesiologist's concept', in H. Delooz (ed.), *Emergency Medicine and the Anaesthetist* Vol 6, Baillière Tindall, London-Philadelphia-Sydney-Tokyo-Toronto, p. 1–23.

Dolenc, D.A., et al.: 1985, 'DRGs: the counterrevolution in financing health care', *Hastings Center Report* June, 19–29.

Engelhardt, H.T., Jr, et al.: 1986, 'Intensive care units, scarce resources and conflicting principles of justice', *JAMA* 255, 1159–1164.

Gregory, C.L.: 1984, 'Ethics: a management tool? A profile of the values of hospital administrators', *Hospital & Health Services Administration* March/April, 102–119.

Jennett, B.: 1991, 'Diagnosis and management of head trauma', *Journal of Neurotrauma* 8 suppl., S15–S19.

King's Fund Panel: 1989, 'Report on intensive care in the United Kingdom', *Anaesthesia*, 428–431.

Knaus, W.A.: 1991, 'Short-term mortality predictions for critically ill hospitalised adults: science and ethics', *Science* 254, 389–394.

Knaus W.A.: 1991, 'The Apache III prognostic system. Risk prediction of hospital mortality for critically ill hospitalised adults', *Chest* 100, 1619–1635.

Miller, C.F., et al: 1984, 'Utilization of arterial blood gas determinations in surgical intensive care unit patients', *Critical Care Medicine* 12, 318.

Morreim, E.H.: 1985, 'The MD and the DRG', *Hastings Center Report* June, 30–38.

Morris, A.H.: 1993, 'Paradigms in management', in R. Pinsky and J.F. Dhainaut (eds), *Pathophysiologic foundations of critical care*, Williams & Wilkins, Baltimore-Hong Kong-London-Munich-Philadelphia-Sydney-Tokyo, p. 193–206.

Owen, D.: 1984, 'Medicine, morality and the market', *Canadian Medical Association Journal* 130, 1341–1345.

Pace, N.L.: 1985, 'But what does monitoring do to patient outcome?', *International Journal for Clinical Monitoring and Computing* 1, 197–200.

Richards, G.: 1984, 'Technology costs and rationing issues', *Hospitals* June 1, 80–86.

Ruark, J.E., et al: 1988, 'Initiating and withdrawing life support. Principles and practice in adult medicine', *New England Journal of Medicine* 318, 25–30.

Schneiderman, L.J., et al: 1990, 'Medical futility: its meaning and ethical implications', *Annals of Internal Medicine* 112, 949–954.

Spivey, B.E.: 1984, 'The relationship between hospital management and medical staff under a prospective payment system', *New England Journal of Medicine* 310, 984–986.

Stein, J.: 1985, 'Industry's new bottom line on health care costs: is less better '?, *Hastings Center Report* October, 14–18.

Task Force on Ethics of the Society of Critical Care Medicine: 1990, 'Consensus report on the ethics of foregoing life-sustaining treatments in the critically ill', *Critical Care Medicine* 18, 1435–1439.

Thomasma, D.C.: 1985, 'Hospital ethics committees and hospital policy', *Quality Review Bulletin* July, 204–209.

Truog, R.D. et al: 1992, 'Sounding board. The problem with futility', *New England Journal of Medicine* 326, 1560–1564.

Veatch, R.M.: 1981, *A Theory of Medical Ethics*, Basic Books Inc., New York.

Veatch, R.M.: 1985, 'The HMO physician's duty to cut costs', *Hastings Center Report* August, 13–14.

Walker, R.M.: 1991, 'DNR in the OR. Resuscitation as an operative risk', *JAMA* 266, 2407–2433.

Webster, G.C.: 1991, 'Evaluation of a "do not resuscitate" policy in intensive care', *Canadian Journal of Anaesthesiology* 38, 553–563.

PART TWO

MORAL REFLECTIONS ON THE CLINIC

KEVIN WM. WILDES, S.J.

CONSERVING LIFE AND CONSERVING MEANS: LEAD US NOT INTO TEMPTATION

Human finitude is inescapable. The allocation and use of medical resources, particularly those of critical care medicine, brings us face to face with human finitude in at least two ways. First, we are confronted with the limits of our resources. A society simply cannot provide all beneficial health care that is possible to those in need. Each choice to allocate resources in one area is a choice to limit resources in another. The resources expended on a macro-level for health care means that there will be less to spend on other areas of social need such as education. Within medicine itself the same principle of choice applies. For example, the resources devoted to critical care medicine are resources that will not be devoted to primary care or some other area of medicine. These examples capture a sense of the necessity, in the face of human finitude, to balance competing goods.

Critical care medicine confronts us with our finitude in a second way. We will all die. While the limit of death touches all human beings there is a wide spectrum of reactions to death. Some may seek to end life as soon as it becomes burdensome or painful while other will seek to prolong life at all costs. Thc attcmpt to avoid death at all costs is costly for others as it drains resources from them. This attempt exacts moral costs as well.

The Christian response to death is radically different from either of these two extremes. In a Christian perspective, there is a basic obligation to conserve one's life yet death is not to be avoided at all costs (Cronin). The moral question becomes how one is to balance limited resources, the inevitability of death, and the obligation to conserve one's life while being faithful to other moral obligations.

The Roman Catholic tradition has struggled with the question of discerning the limits of using medical technology to preserve one's life. Within the Catholic tradition there is a set of reflections on the use of medical technology which dates to the 16th century. Central to this reflection has been the distinction between ordinary and extraordi-

K. Wm. Wildes (ed.), Critical Choices and Critical Care, 105–118.

nary means. The distinction is conceived as a way to guide patients in fulfilling their obligations to conserve life. In exploring contemporary questions regarding the allocation of health care resources generally, and critical care resources in particular, the distinction can still be helpful not only in dealing with the individual cases but in designing general institutional policy. Health care providers (physicians, institutions, governments) must make choices about the allocation of resources. The ordinary-extraordinary distinction can be helpful in these choices since, at its core, the distinction guides men and women in ordering different goods in their lives. This insight can help provide effect policies which assist patients in leading in leading well ordered Christian lives.

I. RECLAIMING A TRADITION

Advances in medical technology challenge moral reflection in at least two ways. First, technologies not only improve what medicine is able to do and allow it to perform more efficiently but they also allow medicine to do what was once, at best, only imaginable. From conception to death health care has been transformed by the advent of what new technologies can do. The moral challenge is to interpret technological capacities within a moral narrative that the Church has maintained. Such an interpretation enables men and women to decide when technologies are appropriately or inappropriately deployed. Secondly, medical technologies have both human and economic costs. Critical care technology can only be used at considerable expense (See, Brody; Rie; DeLooz, Serani-Merlo and Rosso). The expense of such technology prompts not mere economic questions but moral questions about the stewardship of limited resources and one's obligation to use such resources. Each choice to deploy medical technology is a choice to use our resources in one way rather than another. Each positive determination is a negation. Thus how we determine the "best" use of our resources is a moral choice in that it is determined, in part, by the value commitments we hold. Such technology also exacts human costs in that such technology involves pain, suffering, deprivation, and lost opportunities. These costs, in addition to concerns about the stewardship of resources, raise issues of why and how far we should pursue the conservation of life.

These questions about the uses and limits of medical technology are not new to the Roman Catholic tradition. The sixteenth and seventeenth centuries were an era of tremendous medical change and dramatic advances. It was in the sixteenth century that Vesalius published *De humani corporis fabrica* (1543) from which anatomy, as we know it, emerged (Vesalius). It was the age when Harvey (1578–1657) advanced the theory of circulation and Thomas Sydenham (1624–1689) published *Observationes medicare* (1676) which was a systematic description of diseases. In this Renaissance period one finds medicine, like other disciplines, adopting a methodology of disciplined observation and testing.[1] Medical progress was such that Descartes thought he could live to be 100.[2] But these developments in medicine were not without costs. Many of the therapeutic methods were costly in both monetary and human terms (e.g., surgery without anesthesia).

For the Christian such advances posed a moral dilemma. Charged with being a good steward of the life given to them by God the question arose as to whether there are limits to the obligation to accept life prolonging treatment. That is, confronted with new technological developments in medical treatment (e.g., surgery) was one obliged to pursue every possible treatment in light of the natural law duty to preserve life. It would seem obvious that one did not have an obligation to pursue treatments that would not be beneficial. The question which was more difficult was whether or not one could refuse treatment even when it held out the possibility of restoring health.

The tradition responded by arguing that a person was not obligated to pursue all life prolonging therapies. Roman Catholic theologians tried to draw a distinction between obligatory and non-obligatory treatment and this distinction was cast in terms of ordinary and extraordinary means.[3] The distinction not only helped to answer difficult, practical questions about particular treatments but it also helped to conceptualize an understanding of human life and a Roman Catholic's spiritual disposition towards life, health, and death.

Now, three centuries later, the distinction still has moral relevance and conceptual vitality. The spiritual and moral sensibilities encompasses in this distinction are just as valid today as they were three centuries ago. However, this distinction, like any other concept, runs the risk of becoming ambiguous in meaning if it is taken out of context. The life of a community, embodied in its traditions of thought and action, can be easily lost. The intellectual and moral life of community must be rediscovered anew each time it is confronted with new challenges (MacIntyre). As in the intellectual life of any tradition, the historical and conceptual framework for the distinction of ordinary and extraordinary means needs to be set out. Then, one needs to explore how this tradition can respond to contemporary issues of resource allocation in critical care medicine.

II. THE FRAMEWORK OF THE OBLIGATION

The obligation to sustain one's life is rooted in a particular moral view. Indeed, it may strike many people, in a morally secular society, as odd that one would have an *obligation* to sustain one's life. Instead others may argue that remaining alive is a good which, when it becomes too burdensome, should be ended. At the same time many, in the wake of debates about abortion, have taken the view that human life is an absolute value which should be pursued at all costs. The Roman Catholic tradition, by understanding the obligation to conserve one's life as an imperfect duty, is one that has moved between the extreme attitudes of disregard and vitalism.

In his doctoral dissertation which examined the obligation to conserve human life, Bishop Daniel Cronin, points out that man lack perfect dominion over his life. Life is a gift from God over which man has an "important dominion" (Cronin, p. 10).[4] Man's dominion over his life is different from the dominion he has over the rest of the creation. Cardinal de Lugo (d. 1660) argues that there is a fundamental difference between the dominion man receives over things and the dominion he receives over his own life.

> Now we prove that man is not the master of his life this way: although man can receive dominion over things which are extrinsic to himself or which are distinct from him, he cannot, however, receive dominion over himself, because from the very concept and definition, it is clear that a master is something relative, for example, a father or a teacher; and just no one can be father or teacher of himself, so neither can he be master of himself, for to be master always denotes superiority with regard to the one over whom he is the master. Hence, God himself cannot be master of Himself, even though He possesses Himself most perfectly. Therefore man cannot be master of himself, however, he can be master of his operations, and therefore, he can sell himself and thus, improperly speaking, we might say he gives mastery of himself to another but, he really does not give over mastery of himself basically or radically, but only mastery over certain of his operations,...therefore, a man can dispose of his own operations of which he is master, not of himself, (or to say the same thing) not of his own life over which he is not master, nor can he be (de Lugo, Disp. X, Section I, no. 9).

What emerges in the Roman Catholic tradition is an understanding of human life in which there is great, though limited, freedom given to human beings.

One limitation is that life is something not wholly one's own of which to dispose. There is an obligation to conserve human life. This obligation entails that a person not take his or her own life (Aquinas II, II, q. 64. art. 5). This obligation is expressed, for example, in the prohibition against suicide. The more difficult dimension of the obligation, however, is to determine the positive duty one has to conserve life. What is the nature and extent of one's duty to preserve life? This question, of course, has particular importance in health care is determining the extent to which a man or woman must seek treatment for illness. In thinking through a person's obligation to seek treatment the tradition has set out a distinction framed in terms of an understanding of one's earthly life and ultimate end. Life is understood as a gift in relationship to a person's salvation and eternal happiness. So the tradition has argued for an obligation to conserve life that is balanced by other obligations and by the view that it is the love of God that orders all obligations.

Life then is the gift which has both conditions and boundaries. It is a gift which ought not to be deliberately destroyed (e.g. suicide, intentional killing). It is also a gift which brings the duties of stewardship to care for one's health. Indeed in Roman Catholic social teaching one finds these themes repeated. Pope Leo XIII argued, in *Rerum Novarum*, that there was a duty, common to all, to remain alive and from this duty came the right of acquiring the goods by which life is sustained (Denziger, no. 1938). This duty to conserve life can only be fully understood in light of one's salvation. On the one hand man has a duty to maintain the gift that has been given while on the other hand this duty is not absolute and must be seen in light of one's eternal life.

III. THE ELEMENTS OF THE DISTINCTION

The demand to conserve one's life is an imperfect duty and not an absolute one. The natural law tradition has commonly distinguished between affirmative and negative precepts of the law (Noldin-Schmidt, Vol. 1 p. 179, #177). Negative precepts are always binding since they forbid what is intrinsically evil. Death itself would not even make such the performance of such acts licit. They bind *pro semper*. Affirmative precepts bind individually *semper* in that man is not bound, always and everywhere under every circumstance, to do something positively good. He is not bound always and everywhere to fulfill an affirmative precept. An affirmative duty is not a duty which must be carried out when one is faced with physical or moral impossibilities (e.g., Kelly, pp. 8–9). In the case when an individual is unable to make use of the means of conserving his life such a situation presents a physical impossibility. A person nonetheless may physically be able to utilize means for preserving life but still be incapable, because of fear, danger, or inconvenience, of using the means available then such would be a moral impossibility have been called extraordinary. While those which did not were called ordinary means (diposito ad finem per media proportionata) (Merkelbach, I. n. 222). The distinction between the two is not exact. The tradition elaborates certain elements which are central to the description for distinguishing ordinary-extraordinary means.

Before setting out these elements it is important to point out two threads which run through the different elements in the description. First in each element of the distinction the question of a norm for the element arises. Discussion in the tradition has often explored the question of whether or not there is an absolute or relative standard for each element (e.g., expense). In general the writers have opted for the relative standard in the discussion of each element. The second methodological point concerns the importance of *prudent* judgment. Prudence, in contemporary philosophy, has come to mean the notion of rational self interest. However, within the Roman Catholic tradition prudence has been a cardinal virtue which judges not only according to rational self interest but according to the presence of, and response to, the Holy Spirit.

The medieval understanding of prudence was grounded in the Aristotelian virtue of *phronesis* which was an intellectual virtue required by the other virtues (Aristotle, NE 6.13, 1144 b 17–1145 a 6; 10.8 1178 a 16–19). It was not moral wisdom but the ability to be practically wise. Phronesis the exercise of right choice in particular case, in the light of moral universal knowledge. Cicero translated the Aristotelian term *phronesis* as *prudentia* and spoke of it in moral terms (Cicero, I, 43). For St. Thomas prudentia was one of the four cardinal virtues. Prudence was the knowledge of what should be done and avoided and it guided reasoning about what ought to be done (Aquinas, I-II, Q. 57, a. 4–5; q. 61, a. 2; II-II, q. 47, a. 5). The prudent man, in the world of medieval casuistry, had a Christian sense about how one ought to act. This moral sense shaped his judgment. Lacking absolute standards of ordinary and extraordinary care, the tradition has relied on prudential judgment: the judgment which seeks to discern God's will. The assessment of whether or not a means is ordinary or extraordinary will depend on the prudent judgment of the individual.

Extraordinary Means are those that would constitute a moral impossibility for human beings in general (absolute norm) or for a particular person (relative norm). Cronin points out that theologians have used a variety of phrases to describe such impossibilities. However, following his lead, one can gather these elements together into five general headings.

1. *Quaedam impossibilitis*. The factor of a *certain impossibility* exists in any case when it is either impossible to obtain the means or in using them even if they were attainable. As Kelly points out, however, "[I]n concrete cases it is not always easy to determine when a given procedure is an *extraordinary* means. It is not computed according to a mathematical formula, but according to the reasonable judgments of prudent and conscientious men" (Kelly, 1958, p. 135). The essential element of an extraordinary means is that they present a moral or physical impossibility.

2. *Summus labor and Nimis dura*. Roman Catholic moralists have relied on expressions such as *greatest effort* or *too difficult* to describe when a means require excessive effort. A favorite example of excessive effort often used in the tradition has been that of the requirement of a long journey to live in a more healthful land. Following the tradition Zalba, more recently, has argued that one is not bound to submit himself to a very dangerous operation or to a burdensome convalescence (Regatillo and Zalba, Vol. II, p. 269).

3. *Quidam cruciatus and Ingens dolor*. The question of pain is an element which comes quickly to mind when deciding if a means is extraordinary or not. A common example in the tradition is the discussion of amputation. The distinction originated at a time when surgery, without anesthesia, was practiced. While advances in medicine allow us to control and alleviate pain to a large extent, considerable pain and discomfort are involved in many procedures. One can think as a contemporary example of the pain and discomfort involved in chemotherapy for oncology patients.

4. *Sumptus extraordinarius, Media pretiosa and Media exquisitat.* The moral theologians have always taken into consideration the element of expense. Expense can constitute an excessive hardship which, in turn, can make the means extraordinary. When the moralists in the tradition have spoken of cost they have done so in terms of the expression *secundum proportionem status*. The norm of excessive expense is relative to one's economic status. It also varies country to country (Cronin, p. 108). Medical expense must be considered in light of a person's financial position.

This element poses an important point of reflection for contemporary Roman Catholic thought on health care. The tradition has recognized different stations and economic inequalities as part of the human condition. Thus, what could be considered extraordinary for one person may be ordinary for another. However, much of contemporary discussion, both secular and religious, has focused on equality of treatment in health care. Contemporary Roman Catholic social teaching will have to address its older, medical-moral tradition.

5. *Vehemens horror*. Moralists in the tradition have attended to the emotions of fear and repugnance. When fear is present, even when judged to be irrational and unwarranted, it can make a means extraordinary. Distaste for a particular treatment can render a means extraordinary (e.g., being treated by a physician of the opposite sex; the repugnance of the mutilation or amputation).

In each of these five elements Cronin points out that the reaction of the patient (e.g. fear) is what constitutes the extraordinary means. It is not that the extraordinary nature of the treatment lies within the treatment itself. Rather, it is in the patient's reaction to the treatment, the burden the treatment imposes on him or on others.

Ordinary Means: The ordinary means of conserving one's life are to be distinguished from ordinary medical procedures in that what may be an ordinary (common) medical procedure may be nonetheless extraordinary to a patient. In understanding what is an ordinary means of conserving one's life the tradition has outlined several elements.

1. *Spes salutis*. An ordinary means of conserving life must offer some hope of beneficial result if it is to be ordinary and considered obligatory. The hope of benefit must be more than simply the hope of postponing the inevitable. De Lugo gives an example of a man condemned to death by fire (*De Justitia*, Disp. X, Section I, no. 30). In the case, de Lugo supposes that while surrounded by flames the man notices that he has sufficient

water to extinguish some of the fire, but not all of it. He then asks if the man is obliged to use the water and he answers "No". Even though the means is common and it would offer some benefit (prolonged life) it is considered of no value since it would only postpone the inevitable. The benefit derived must be worthwhile in both "quality and duration".

2. *Media communia.* Another element frequently mentioned is that the means must be common. However, in determining whether a means is common it is necessary to consider the relative factors involved (Cronin, p. 93).

3. *Secundum proportionem status.* The social position or particular status are relevant factors in deciding if a means is common or not. These elements are part of the evaluation of the cost of a treatment. De Lugo helpfully discusses this element with the case of a novice in religious life who is suffering from ill health. The novice may be told that the best treatment of his illness is to return to the world in order to have the food and surroundings necessary for his recovery. He argues that the obligation to conserve one's life does not require the novice to return to the world and that he satisfies his obligation by using the means which other men in the religious life commonly use (*De Justitia*, Disp. X, Section I, no. 36).

Again the tradition accepts different social strata and "entitlements" in determining the ordinary-extraordinary distinction.

4. *Media non difficilia.* The notion of "difficulty" is generic in the tradition according to Cronin (p. 96). That is, the notion is general and not specific. It can involve elements of pain, cost, danger to life, fear, etc. In making a judgment about the difficulty of a treatment the tradition tries to balance the serious demand of the natural law to conserve one's life with the proportionate difficulty of fulfilling the law.

5. *Media facilia.* Ordinary means are those which can be employed conveniently and reasonably. Ordinary means are reasonable means and they can entail moderate difficulty.

Summary. The obligation to sustain one's life extends to using those means which do not cause and undue burden or moral impossibility and which offer some reasonable expectation of health. This obligation is rooted in a faith that one's life is not one's own. The core of the distinction is not merely a question about determining the obligation one has towards preserving one's life. Rather, there is a spiritual insight that "life" is not the paramount value (de Lugo; Pius XII). De Lugo rules out the necessity of an extraordinary means because the "bonum" of a man's life is not so important as to demand conservation by any and all possible means. De Lugo writes:

> ...the bonum of his life is not of such great moment, however, that its conservation must be effected with extraordinary diligence: it is one thing not to neglect and rashly throw it away,...it is another however, to seek after it and retain it by exquisite means as it is escaping away from him...(*De Justitia et Jure*, Disp. 10, Sect. 1)

Pope Pius XII spoke of the good of life as subordinate to "higher spiritual ends" (p. 212). In the view of the tradition the conservation of life is an imperfect obligation which must be ordered in relationship to other obligations. If one gives the "life" improper weight one runs the risk of idolatry. Preserving one's life is an imperfect obligation ordered to attaining a higher spiritual end.

One finds the concerns expressed by the distinction expressed in contemporary Roman Catholic documents. The 1980 *Declaration on Euthanasia* expresses these same concerns (Congregation). The document speaks of the "duty to care for his or her own life or to seek such care from others" (p. 222). Yet the document speaks of the limits on this obligation. One is not obliged to seek experimental treatment and can "make due with the normal means that medicine can offer" (p. 223). It also speaks of situations "[w]hen the inevitable death is imminent in spite of the means used, it is permitted in conscience to take the decision to refuse forms of treatment that would only secure a precarious and burdensome prolongation of life..". (p. 223). One needs to judge the means used by "studying the type of treatment to be used, its degree of complexity or risk, its cost and the possibilities of using it, and comparing these elements with the result that can be expected, taking into account the state of the sick person and his or her physical and moral resources" (p. 222). Finally, the Congregation captures the essence of the tradition when it writes: "Life is a gift of God, and on the other hand death is unavoidable; it is necessary, therefore, that we, without in any way hastening the hour of death, should be able to accept it with full responsibility and dignity" (p. 223).

IV. THE ALLOCATION OF RESOURCES

It is not immediately evident how this distinction can help us to think through the allocation of critical care resources. The distinction between ordinary-extraordinary means has, historically, centered on the duty of patients to preserve their lives. It is clearly helpful in assisting patients, and their families, sort through their choices about the uses of such medical technology. For while such technology is often part of the ordinary practice of medicine it should not be assumed that such technology is part of the ordinary duty one has in conserving one's life. Such care may offer little benefit and great burden to a patient.

How can health care providers deploy the distinction of ordinary and extraordinary means to think about the use of their resources? At first glance it would not seem very helpful for policy development as it is patient centered. In the Roman Catholic tradition it would seem that the allocation of resources is best addressed by the social teaching of the tradition and a thematic understanding of stewardship.

However there are at least three ways in which the distinction can be helpful beyond the individual perspective in framing policy in critical care medicine. One element in the tradition for determining the morally burdensome nature of a treatment was not only its affects on a patient but also its effects on others (Pius XII; Congregation). In contemporary medicine an important change has occurred in that the resources for health care are situated in a network of public and private financial support. It would seem that one can begin with the notion of burden to the family, found in the tradition, and extend it to the burden on others in the fiscal network since they are sharing in the cost of care. So, while treatment may be common, and even hold out the hope of recovery, it may still be extraordinary as it imposes a grave burden on others. It would seem that members of the network may refuse to commit resources to bear this burden.

There is yet another way in which the distinction may be helpful to Roman Catholic health care institutions and providers. One can capture the distinction by arguing that it helped an individual order his/her obligations and commitments. The distinction reminds Catholics that human life is ordered toward higher spiritual ends (Pius XII). Indeed ordinary means dispose one to the *end* of human life through proportionate means. It is possible that the resources of critical care medicine, or any medical service, may pose a real moral and spiritual threat to the patient and the patient's family and friends in that it may be lead one into a disordered spiritual life. The heart of the distinction there is the obligation to preserve one's life, however, this obligation is set in the context of other obligations. The inappropriate use of critical care medicine may lead to a disordered set of values in that it can create an idolatry of life and forget that this life reaches consummation in "higher spiritual ends" (Pius XII). The virtue exhibited in preserving one's life may be turned into a vice (Bole).

In light of this tradition Roman Catholic health care institutions may well establish policies of limitation made in the spirit of this distinction. The justification for limiting such resources may be, not only and argument from stewardship, but an argument that making such resources available may well lead the patient, or the patient's family, into the idolatry of human life. If this were to happen the institution would have participated in a moral evil by putting the life of the patient before all else as one worships at the altar of critical care medicine.

A third element in the distinction of the ordinary and extraordinary means, which may be helpful to questions of allocation, is the emphasis in the tradition on the extraordinary nature of treatments which offered no hope of benefit to patients. Cronin concludes that the tradition emphasizes that a treatment must offer some *hope of beneficial result* if it is to be considered obligatory (p. 143). A fourth, and related element, is that the tradition does not view a treatment as obligatory when it offers only a brief prolongation of life. De Lugo's example of the burning man with the water illustrates this point. The man is not obligated to use the water since it will not extinguish the flames and allow him to escape death. A fifth point from the tradition is that it allows for variation in the treatment of patients. What is ordinary-extraordinary will vary according to one's station in life. The tradition of the distinction certainly does not endorse a simplistic notion of "equality" in the prolongation of life.

CONCLUSIONS

The deployment of critical care technology and the resources to support it provoke powerful moral questions about the allocation of resources and the obligation to sustain life. These questions are not entirely new to the Catholic tradition since the sixteenth century has reflected an obligation to conserve life. The distinction of ordinary and extraordinary means provides a way to balance the duty to conserve one's life and the burdens of treatments on a patient and others. The distinction helps to order obligations towards others and in doing so formulate institutional and professional policy. Imperfect duties of beneficence are a judgment of the outcome in relationship to the cost. One might formulate the judgment in terms of a formula:

$$\text{Duty} = \frac{\text{Quality of Outcome}}{\text{Cost}}$$

The distinction can guide the development of institutional policy in that it must account for the cost of treatments in terms of the burdens, both spiritual and economic, to the patient and others.

The distinction also reaffirms a basic view of human life that it is not to be disposed of at our own will nor is it to be idolized. It provides a challenge for Catholic health care institutions and professionals to think through their own practice of medicine. Hospitals or physicians may well lead a patient or their family into the temptation of idolizing human life which is a temptation to be avoided.

NOTES

1 The influence of Bacon on Sydenham, for example, is seen in the later's writings: the improvement of phyfick in my opinion depends (1) upon collecting as genuine and natural a defcription, or hiftoyr of all difeafes, as can be procured and (2) layin gdown a fix'd and complete method of cure. It is eay enough to defcribe difeafes unfkilfully, but to write fuch a full and accurate hiftory of them as to efcape the cenfure that lofe *Bacon* has paffed upon from great promifers in another way, is much more difficult taft...An thus the cure of difeafes is eafily delivered according to the common method, but to do it in a mafterly fcientifcal manner will appear a mucch harder tafk to thofe who know that there are abundance of diftempers to be met with in practical writers, that were ftill continue incurable (Sydenham, "The Author's Preface", pp. ii–iii.

2 *The Antwerp Extra ordinisse Postijdinghe* of April 10, 1650 reported that "...in Sweden a fool has died who claimed to be able to live as long as he wanted" (Lindeboom, p. 94).

3 One of the first known theologians to begin to working out a distinction between those means of conserving one's life which were obligatory (ordinary) and those which were not obligatory (extraordinary) was F. Victoria (d. 1546). In his reflections on homicide, suicide, and the virtue of temperance he examines a number of cases where what might appear to as obligatory (e.g. eating) could be considered non-obligatory (See, D. Gracia, this volume).

4 In the traditional discussions of suicide theologians have pointed out that man's dominion over the created world, and his own life, is a dominion given by God. So it is God alone who has supreme and ultimate power over it (See, Deut. 32:39). St. Thomas argues that man is not the master of his own life (Aquinas, II, II, q. 64, art. 5).

REFERENCES

Aquinas, T.: 1948, *Summa Theologica*, Christian Classics, Westminster, Maryland.

Aristotle: 1984, *Nicomachean Ethics*, in *The Complete Works of Aristotle*, J. Barnes (ed.), Princeton University Press, Princeton, N.J.
Brody, B.A.: 1995, 'Ethical issues raised by intensive care', this volume, 7–22.
Congregation For the Doctrine of the Faith: 1992, "Declaration on Euthanasia", in *Birth, Suffering, and Death: Catholic Perspectives at the Edges of Life*, K. Wildes, S.J., Abel F., S.J., Harvey, J.C., (eds.), Kluwer Academic Publishers, Dordrecht, The Netherlands.
Cronin, D.A.: 1989, *Conserving Human Life*, Pope John XXIII Medical-Moral Research and Educational Center, Braintree, Massachusetts.
DeLooz, H.: 1995, 'Ethical issues in critical care: Criteria for treatment', this volume, 79–101.
de Lugo, J.: 1869, *De Justitia et Jure*, Vives, Paris.
de Lugo, J.: 1869, *Disputationes Scholasticae et Morales*, Vives, Paris.
Denziger, H., Schonmetzer, A.: 1963, *Enchiridion Symbolorum*, Herder, Freiburg im Breisgau.
Gracia, D.: 1995, 'The old and new in the doctrine of the ordinary and extraordinary means', this volume, 119–125.
Kelly, G.: 1950, "The duty of using artificial means of preserving life", *Theological Studies* XI, pp. 203–220.
Kelly, G.: 1958, *Medico-Moral Problems*, The Catholic Hospital Association, St. Louis.
Lindeboom, A.B: 1978, *Descartes and Medicine*, Rodopi, Amsterdam.
MacIntyre, A.: *Whose Justice? Which Rationality?*, University of Notre Dame Press, Notre Dame, Indiana.
Merkelbach, B.: 1935, *Summa Theologiae Moralis*, Desclee de Brouwer et Soc., Paris.
Noldin, S.J., H. and Schmitt, S.J., A.: 1940, *Summa Theologiae Moralis*, Feliciani Rauch, Rome.
Pius XII: 1992, "The Prolongation of Life", in *Birth, Suffering, and Death: Catholic Perspectives at the Edges of Life*, K. Wildes, S.J., F. Abel, S.J., J. Harvey, (eds.), Kluwer Academic Publishers, Dordrecht, The Netherlands.
Regatilla, E., M. Zalba: 1952, *Theologiae Moralis Summa*, Biblioteca de Autores Cristianos, Matriti.
Rie, M.A.: 1995, 'Ethical issues in intensive care: Criteria for treatment', this volume, 23–56.
Serani-Merlo, A. and Rosso, P.: 1995, 'Intensive care: Its nature and problems', this volume, 63–78.
Sydenham, T.: 1753, *The Entire Works of Dr. Thomas Sydenham*, John Swan (ed.), E. Cave, London.
Vesalius, A.: 1543, *De Humani corporis fabrica librorum Epitome*, J. Opernis, Basil.
Victoria, F.: 1587, *Reflectio IX – De Temperantia*, Lugduni.

DIEGO GRACIA

THE OLD AND THE NEW IN THE DOCTRINE OF THE ORDINARY AND THE EXTRAORDINARY MEANS

INTRODUCTION

The essay by Kevin Wildes "Conserving Life and Conserving Means: Lead us not into Temptation" is a good explanation of the classical theory of the ordinary and the extraordinary means. It is not frequent to find a correct explanation of this doctrine. If we take a look at the classical manuals of moral theology, we can realize that this doctrine has been more used than analysed; in other words, that it has been understood in a way more intuitive and emotional than properly rational. Only with the classical studies of Kelly (Kelly, 1950) and Cronin (Cronin, 1958) the history of this theory began to be correctly known. One of the virtues of the paper of Wildes is the use he makes of these classical studies.

I. SALAMANCA: BETWEEN NOMINALIS AND THOMISM

The conclusions of the studies of Kelly, Cronin, McCormick (McCormick, 1985), Paris (Paris, 1992) and Wildes have a very peculiar meaning for the Spanish ethicists, and especially for one who studied Philosophy and Medicine in Salamanca the city and the University in which, in the Sixteenth century, the doctrine of extraordinary means was built by a professor, Franciscan by name and Dominican by order, Francisco de Vitoria, and two other who were Dominicans by name and also by Order, Domingo de Soto and Domingo Bañez. Perhaps it is not completely superfluous remembering briefly the climate of the intellectual revolution promoted by Vitoria and Soto, in order to understand the real meaning of the theory we are talking about.

The distinction between ordinary and extraordinary means appeared in the School of Salamanca of the 16th century, when Domingo de Soto affirmed that the religious superiors could only require their subjects to use medicine that could be taken without too much difficulty (Paris, 1992, p. 199). In 1587, the Dominican moralist Francisco de Vitoria

K. Wm. Wildes (ed.), Critical Choices and Critical Care, 119–125.

wrote in his *Relectiones theologicae* (Vitoria, 1557) a commentary on the obligation to use food to preserve life. He asked: "Would a sick person who does not eat because of some disgust for food be guilty of a sin equivalent to suicide"?. And he replied: "If the patient is so depressed or has lost his appetite so that it is only with the greatest effort that he can eat food, this right away ought to be reckoned as creating a kind of impossibility, and the patient is excused, at least from moral sin, especially if there is little or no hope of life". Vitoria provides an everyday example of the type of "delicate treatment" that would be beyond what one is obliged to employ to preserve life: "Chickens and partridges, even if ordered by the doctor, need not be chosen over eggs and other common items, even if the individual knew for certain that he could live another 20 years by eating such special foods". The final stage of this movement was Domingo Bañez (McCartney, 1980), who proposed in 1595 the distinction between "extraordinary" and "ordinary" means, by which was meant measures proportionate to one's condition or state in life. Thus, if something were very costly or burdensome or if it did not offer substantial benefit to the patient, there was no moral obligation to use it. This standard applied even to lifesaving measures (Paris, 1992, p. 199).

All those men, Vitoria, Soto, Bañez, were theologians nourished in Nominalistic philosophy and theology, but they were also aware of the intellectual power and the possibilities of Thomistic philosophy. All the novelties introduced by the members of Salamanca's School in the field of moral theology, were consequence of the reinterpretation of the Aristotelian and Thomistic tradition from the point of view proper of the Nominalistic tradition (Andrés, 1977, Vol. 2, p. 80).

Nominalism was particularly important in moral theology. Manuals of moral theology often argued that nominalism introduced casuistry, the concrete and singular analysis of moral problems, instead of their abstract and principialistic approach. But this is not completely true. What nominalism attempted was a new formulation of the Aristotelian work, emphasising their concrete and individual dimensions and leaving the universal and abstract in a second place. From the nominalistic point of view, some Aristotelian works, like *Nicomachean Ethics*, *Politics* and the biological works took priority over other works, particularly Aristotle's *Metaphysics*. Without this change of perspective, the intellectual revolution of the Sixteenth Century remained inexplicable. The importance that the *Politics*, *Ethics* and in general the practical, manual, artistic and interested aspects of life, reached against the more theoretical and disinterested, was probably the most distinctive peculiarity of the so-called 'Humanism' of the age. Humanism was not only the consequence of the discovery of printing by Guttenberg, and the easy possibility of reading old classical books. It was also the outcome of a particular point of view, different from the medieval and scholastic. If great scholastics emphasised the more platonic aspects of the Aristotelian work, now the most pragmatic and concrete aspects of the Aristotelian books took shape. The consequence was a new reading of those works, perhaps more accurate and coherent than the traditional.

The generation of Vitoria and Soto was educated in nominalism, but also, as black friars, in the philosophy of Thomas Aquinas. They attempted to "aquinaze" nominalism, and "nominalize" aquinism. This means that they tried to interpret the theology of Thomas Aquinas and Aristotle's philosophy in a modern way, emphasising the concrete and individual aspects of both works.

II. THE NATURALISTIC ORIGIN OF THE DOCTRINES OF PROBABILISM AND EXTRAORDINARY MEANS

Three characteristic principles of Catholic moral theology, the principle of double effect, the distinction between ordinary and extraordinary means, and probabilism, took shape in this milieu. Two of them, began in the Salamanca in the second half of the Sixteenth century. Probabilism was conceived in first place by another black friar of Salamanca: Bartholomeo Medina in 1577, when, in his *Summa Theologica*'s commentary (*S.Th.*, 1–2, q.19, a.6, concl.4), he analysed the following question: "If it is necessary to follow the most probable opinion, rejecting the only probable, or if the only probable can be followed"[1] And he answered: "Certainly, some arguments seem to be definitive, but it seems to me that, if an opinion is probable, it is licit to follow it, even though the opposite opinion is more probable".[2]

The meaning of these words of Medina has been frequently discussed from the time of Pascal until now. From the very beginning, many authors interpreted the word probability in the medieval sense of "provable" or capable of being proven. Thomas Aquinas considered moral judgements as having "probable certitude which, for the most part, attains the truth, even though in some instances it fails to do so" (*S.Th.*, 2–2, q.70, a.2; see Jonsen and Toulmin, 1988, p. 165).

Therefore, probability was confused with "moral certitude", i.e. with its direct opposite, "certitude".[3] Only half a century ago, in 1928, M. Gorce (Gorce, 1928) and I. Menéndez Raigada (Menéndez Raigada, 1928) argued that for Medina probability had the meaning of *certum in re contingente*. I do not think so. In my opinion Medina was thinking as a pure Aristotelian, and the meaning of the word probable in his work is the proper of the Aristotelian theory of deliberation and prudence. From an Aristotelian point of view certitude is not possible in particular and contingent judgements (*in re contingente*) but only in universal and necessary (*in re necessaria*). Concrete judgements, like moral ones, can only be probable. Probabilism took shape when philosophers and moralists realized the importance of particular and concrete judgements in Aristotelian work, over the universal and abstract.

This was the atmosphere in which the theory of ordinary versus extraordinary means was built. It was a mere application of the theory that extremely difficult actions can not be considered as morally obligatory. Nature has an internal order, that consists in balance or equilibrium. Therefore, moral virtues are always in the middle (*in medio virtus quando extrema sunt vitiosa*). Because extraordinary actions are not prudent, they can not be obligatory. This is the philosophical foundation of the theory of the extraordinary means.

The problem is that this foundation is naturalistic, and should take the chance of naturalism. This was the reason Protestantism rejected this theory; and after Hume and the discovery of the naturalistic fallacy, the question is if the theory of extraordinary means can have some rational meaning.

III. LOOKING FOR A DIFFERENT FOUNDATION

I think it can have some rational meaning, but different from the traditional. My opinion is that the only reality we can consider absolute in order to set up morality is not nature but human being; a human being that must not be considered as a material or natural reality, but only from a formal perspective. This is an important distinction, because material judgements can not be absolute, and absolute judgements can only be formal (Gracia, 1989, 1991). Formally speaking, man is an end by itself. Man is absolute and must, therefore, be absolutely respected. Here there is no place for exceptions, nor for the theory of extraordinary means. By definition, means are always related with material conditions, not with the form of morality, which is always absolute.

The problem of material judgements is that they can not be absolute, and therefore they always have exceptions. The formal respect due to human beings obliges us not to kill or not to lie to any of them. But killing and lying are material duties, duties with material content, and in its materiality they are not absolute. That is why they have exceptions. For instance, killing as a mean of saving the proper life; or the capital punishment; or in the case of lieing, the so called pious fraud. Exceptions can only be justified, the same as norms, by the formal principle. A norm is the action that is approved by the formal principle. On the contrary, an exception is the action that is in accordance with the formal principle only in some concrete and unusual occasions.

Extraordinary means are exceptions in this way, in view of the consequences. And the problem now is determining what kind of consequences can be considered extraordinary. All? I do not think so. There are, at least, two kinds of consequences: objectives, like price, and subjective, like the different ideals of happiness. In the first case means should be best called proportionate and disproportionate, reserving ordinary and extraordinary only to the exceptions due to subjective values. Proportion is a mathematic relation between two objective quantities, while the language of ordinary and extraordinary only has meaning in the relationship a person has with his own values, beliefs and ideals.

The differences between those two levels are important, because in the first case the decision is generally transitive, and can be made by a person different from that who suffers the consequences. On the contrary, in the second case decisions are intransitive, and can be taken only by the person who will suffer the consequences. Only the own individual can define what is extraordinary for him, but not what is disproportionate. That is why it has no meaning to use the theory of extraordinary means in relation to the ethical principles of non-maleficence and justice. For instance, citizens must defend countries against invaders, putting at serious risk their lives or loosing them, and they can not avoid those obligations saying they are extraordinary acts, or thinking that war is an extraordinary mean. The theory of extraordinary means can be only applied to intransitive actions, those that affect only the subject of the action.

CONCLUSION

To end, I would like to say that the theory of extraordinary means is at least confuse and has a wrong naturalistic foundation. Many people think that this theory can be used without direct reference to its naturalistic foundation. I am not so sure. Using the theory in its classical form, leads necessarily to an unconscious agreement with the naturalistic ethics. And on the contrary, the acceptance of a more accurate foundation leads to the complete reformulation of the theory of extraordinary means. In my opinion, its ambiguity is only a symptom of the real problem: the lack of clear and coherent ethical foundations.

NOTES

[1] Bartolomeo Medina, 1590, q.19, a.6: "Utrum teneamur sequi opinionem probabiliorem relicta probabili, an satis sit sequi opinionem probabilem".

[2] Bartolomeo Medina, 1590, q.19, a.6: "Certe argumenta videntur optima, sed mihi videtur quod si est opinio probabilis, licitum est eum sequi, licet opposita probabilior sit".

[3] This was a problem all over the Middle Age. As Jonsen and Toulmin have written, "novel though it was, Medina's thesis flowed out of broad medieval traditions about the kinds of knowledge required for moral judgement. The theologians and jurisconsults had derived theories about the impact of ignorance on imputability, while the theologians had settled notions about the erroneous conscience. They also recognized the reality of moral doubt, which they defined as the inability of the mind to give assent to either of two contradictory propositions: in such doubt judgment was suspended. They unanimously taught that it was sinful to act while in doubt: to do so was to hold the law in contempt, and one's moral obligation was to resolve the doubt by investigation, prayerful reflection, and recourse to wise and prudent advisors. Should doubt persist, medieval authors unanimously recommended that the decision favor the safest course – that is, the action most likely to avoid sin and to safeguard one's salvation. For example, the medieval theologian, Alexander of Hales, posed the case of a priest who wondered whether a particular contract involving receipt of money for some spiritual benefit constituted the sin of simony, recognizing that some canonists considered it so and that other did not. Alexander answered, 'Being in doubt, he should avoid such a contract, for it is less evil to suffer a temporal loss than a spiritual one'. The moral stance common to medieval theologians seems to have been advocacy of the *via tutior*, the safer way".

"At the same time, the medieval authors also regularly distinguished between doubt and opinion. From both the Ciceronian tradition of 'probable reasons' and the Aristotelian teaching about the nonscientific nature of moral reasoning, they inferred that the certitude of moral judgement was a 'probable certitude which, for the most part, attains the truth, even though in some instances it fails to do so'. Probable certitude was opinion, an assent to one proposition, coupled with the acknowledgment that its opposite might be true. The probability was the quality of a proposition – its 'verosimilitude' – which made it worthy of assent. In this sense only one of two contrary propositions could be probable; this one determined assent and commanded moral action. For the medievals, then, opinion was the field of moral knowledge and decision, and they found nothing strange in asserting that a judgement enjoyed 'probable certitude'" (Jonsen and Toulmin, 1988, pp. 165–166. See also Gardeil, 1911; Richard, 1923).

REFERENCES

Andrés, M.: 1977, *La teología española en el siglo XVI*, 2 vol. Biblioteca de Autores Cristianos, Madrid.

Brisbois, E.: 1936, 'Pour le probabilisme', *Ephemerides Theologiae Louvainensis*,13, 80–94.

Cronin, D.A.: 1958, *The Moral Law in Report to the Ordinary and Extraordinary Means of Conserving Life* (Dissertatio ad Laudem in Facultate Theologica Pontificiae Universitatis Gregorianae), Romae.

Deman, T.: 1928, 'Probabilisme', *Dictionnaire de Théologie Catholique* , Eds.: A. Vacant, E. Mangenot, E. Amann, Paris, Vol. 13, 1, cols. 417–619.

Gardeil, A.: 1911, 'La certitude probable', *Révue des sciences philosophiques et théologiques*, 5, 441–485

Gorce, M.: 1928, 'Medina", *Dictionnaire de Théologie Catholique*, Eds.: A. Vacant, E. Mangenot, E. Amann, Paris, Vol. 10, cols. 481–485.

Gracia, D.: 1989, *Fundamentos de Bioética*, Eudema, Madrid.

Gracia, D.: 1991, *Procedimientos de decisión en ética clínica*, Eudema, Madrid.

Jonsen, A.R., Toulmin, S.: 1988, *The Abuse of Casuistry: A History of Moral Reasoning*, University of California Press, Berkeley.

Kelly, G.: 1950, 'The duty of using artificial means of preserving life', *Theological Studies*, 11, 203–220.

McCartney, J.J.: 1980, 'The development of the doctrine of ordinary and extraordinary means of preserving life in catholic moral theology before the Karen Quinlan Case', *Linacre Quarterly* 47, 215–226.

McCormick, R.A.: 1985, 'Caring or starving? The case of Claire Conroy', *America* 152, 269–273.

Medina, B.: 1590, *Expositio in Primam Secundae Angelici Doctoris D. Thomae Aquinatis*, Venetiis.

Menéndez Raigada, I.: 1928, 'El pseudo-probabilismo de Fray Bartolomé de Medina', *Ciencia Tomista*,37, 35–57.

Paris, J.: 1992, 'Nutrition and Fluids: The Catholic Tradition'. In: Wildes, Abel, Harvey, Eds., *Birth, Suffering and Death: Catholic Perspectives at the Edges of Life*, Kluwer, Dordrecht.

Richard, T.: 1923, 'Théories de la certitude morale', *Révue Thomiste*,28, 157–178, 194–205.

Rolland, M.: 1936, 'La fondament psychologique du probabilisme', *Nouvelle revue théologique*,63, 351–364.

Vitoria, F.: 1557, *Relectiones theologicae*, Lugduni.

PAUL T. SCHOTSMANS

ADMISSION TO AND REMOVAL FROM INTENSIVE CARE: A PERSONALIST APPROACH

INTRODUCTION

Over the past twenty years or so, a consensus has gradually emerged in the Western world concerning the moral and legal appropriateness of forgoing certain medical treatments in some circumstances. This consensus is based first on a recognition that not all treatments which prolong biological life are humanly beneficial to the patient, and, second on the general agreement that there is a moral difference between killing, euthanasia, and allowing to die. In the American legal system, these two views have been combined with the legal concepts of autonomy, privacy and liberty. These views reflect the distinction, between morally ordinary (or mandatory) and morally extraordinary (or optional) means of treatment (see, Wildes) and the distinction between killing and allowing to die. In this essay we hope to develop current understandings regarding the interplay between concepts of ordinary and extraordinary care and the Roman Catholic Church's view regarding justice in the fair distribution of health care resources and apply these to the issue of developing criteria for admission to and removal from intensive care. In all of this, our explorations will focus on the specific problem of what criteria should determine whether a patient is to be admitted to or removed from the benefits of intensive care. Our approach will be personalistic, which means that we will develop a relational structure of the medical profession and thereby avoid an extremely individualism often found in the American approach.

I. CLARIFYING CONCEPTS

A. *Ordinary and Extraordinary Means*

The Roman Catholic tradition has long recognized a distinction between ordinary and extraordinary means of prolonging or preserving life. The

K. Wm. Wildes (ed.), Critical Choices and Critical Care, 127–143.

distinction, at least implicitly, goes back several centuries, though it is most often attributed to Pope Pius XII, who repeated it and stressed it during his pontificate in the 1940's and 1950's (see Wildes, this Volume). According to this tradition, it is never obligatory to make use of medical measures which are morally "extraordinary" in order to preserve life. The distinction between ordinary and extraordinary is a useful one within the Catholic tradition, and many Catholic ethicists are hesitant to abandon it even in the face of criticisms. Critics do have a point, however, when they argue that the words used, are open to misinterpretation if the distinction is understood as a medical one. It is rather a moral distinction, and there are no simple technical or statistical criteria for determining the difference. Means that are usually thought of as medically ordinary (no longer experimental, normal hospital procedures in some cases, not requiring IRB protocol approval) may be morally extraordinary. According to Gerard Kelly, theologians generally considered 'ordinary' means as those procedures which can be obtained and used without great difficulty. The 'extraordinary' means is one 'which involves excessive difficulty by reason of physical pain, repugnance, expense' etc. For him, what a prudent man judges as morally impossible is extraordinary. The factors which make a particular means either ordinary or extraordinary are the reasonable hope of success, the physical condition of the patient, the difficulties involved, etc. According to the particular situation of the patient, an ordinary means can become extraordinary. In such situations, for example a patient in irreversible and terminal coma, 'the artificial means not only need not but should not be used'. It seems that he identifies extraordinary means with the means which do not offer a reasonable benefit.

The "reasonable hope of success" is a relative element, because there are degrees of 'success' and 'hope'. It is this 'reasonable hope of success' which makes a procedure ordinary and hence obligatory or not. He says: "A remedy which includes rather great difficulty, though not moral impossibility, is hardly obligatory unless the hope of success is more probable, whereas a remedy which is easily obtained and used seems obligatory as long as it offers any solid probability of success" (Kelly, 1950, pp. 203–220). Kelly gives more weight to the criterion of 'reasonable hope of success' and to the availability of the remedy without great difficulty. The remedy must be a more probable one. He considers all natural means, such as food, drink, etc. and good nursing, appropriate measures to relieve physical and mental pain as proper treatments, and thereby mandatory. The non-obligatory ones are all artificial means such as life-sustaining machines. He also argues for evaluating each case according to its circumstances, and it is the benefit which is one of the main determining factors to decide whether a treatment is mandatory or not. His almost classical definition can be found in a later article:

> Ordinary means are all medicines, treatments and operations, which offer a reasonable hope of benefit and which can be obtained and used without excessive expense, pain or other inconvenience. Extraordinary means are all medicines, treatments and operations, which cannot be obtained or used without excessive expense, pain, or other inconvenience or which if used would not offer a reasonable hope of benefit (Kelly, 1951, pp. 550–556).

Later we observe a modification for the definition of 'extraordinary' means in his book *Medico Moral Problems*. He incorporated the *'for others'* in the definition:

> Extraordinary means are all medicines, operations, which cannot be obtained or used without excessive expense, pain or other inconvenience for the patient or for others, or which, if used, would not offer a reasonable hope of benefit to the patient (Kelly, 1961, p. 135).

One of the characteristics of Kelly's argument is that he analyses the situation also from the point of view of the relatives and of the physician. In his opinion, for the preservation of life, the duty of the patient, of the family, and of the doctor are coextensive. In instances even when from the point of view of the patient's evaluation a procedure is *per se* extraordinary, but he or she whishes to employ it, then the relatives and the physician are obliged to act according to the interest of the patient. Therefore, there is a difference between the duty of the patient and that of the relatives and of the doctor. Kelly is aware of the fact that in concrete cases it is difficult to determine whether a procedure is extraordinary or not. It is to be decided 'according to the reasonable judgement of prudent and conscientious men'.

B. The Principle of Proportionality

In a document issued by the Congregation for the Doctrine of Faith in 1980, we find references to the concept of ordinary and extraordinary means, when dealing with the problem of euthanasia. Treating on the question of the value of life, the theme of euthanasia, and the meaning of suffering, the document *Declaration on Euthanasia* refers to ordinary and extraordinary means under a special heading 'The Proportionate Use of Therapeutic Agents' (AAS, 1980, 542–552). The document refers to the 'right to die', i.e. the right of man to die peacefully and in a manner worthy of a Christian and a human being. While treating the validity of the principle that guides the concept, ordinary and extraordinary, the document points out that due to the rapid advances of medical treatment and because of its vagueness, today the language of "ordinary" and "extraordinary" appears to be ambiguous. Therefore, the declaration refers to the more recent proposals of some theologians on 'proportionate and disproportionate means': it seems to propose that a morally sound judgment is possible only when there is a proportion between the type of treatment, its degree of difficulty and danger, its expense, the possibility of applying it and the results that can be expected. To find that proportion, and thus to make a choice either for or against the treatment, we have to consider different elements that are relevant, such as the types of the treatment, the complexity and the risk that is involved, the expense, the possibility of applying it, the condition and status of the patient. A further important point which the document mentions is that proportionality is to be decided not only at the beginning of the treatment but also in the course of its application; because during the course of its application the treatment can become disproportionate due to the factors like strain or suffering to the patient, or by being against the wish of the patient, or due to the result.

Hence it is important that we try to briefly look at the principle of proportionality and see how it functions in moral decision making. The general dissatisfaction of the concept of ordinary and extraordinary means for finding a solution to the various problems emerging from the field of medicine (e.g. the situations where good and evil co-exist), led many eminent moral theologians like Janssens, Knauer, Schüller, Fuchs, Van de Poel, Van der Marck, and McCormick to explore a way of reasoning which is now well known as the 'theory of proportionality'. But as noted by Joseph A. Selling, 'proportionate' is neither a 'system' nor a 'determinative methodology', but is is only a way of "looking at things proportionally" (Selling, 1986, p. 165). Louis Janssens in his article *Ontic Evil and Moral Evil* gives the fundamental principles of proportionate reasoning. Though we can examine these different aspects separately, he says, there exists an intrinsic relationship between them. He summarizes them as follows:

> (1) If ontic evil is *per se* intended, the end itself (object of the inner act of the will) is morally bad and, being the formal element (reason and cause of the exterior action), vitiates the entire action... (2) When the single and composite act is viewed from the point of view of reason (secundum rationem), it must be found without an intrinsic contradiction between the *means*(exterior act as material element) and the morally good *end* of the inner act of the will (formal element)... (3) We have the moral obligation to reduce as much as possible the ontic evil which comes about when we act... (4) In the actualization of a good end and the deliberation about the means to this end, the genuinely important question is what place this end has in the totality of human existence... (Janssens, 1979, pp. 69–84).

According to Louis Janssens, proportionality is a question of the relation between end and good. In his book *Ambiguity in Moral Choice*, McCormick gives his own understanding of proportionate reason. For him, proportionate reason means three things: (a)a value at stake at least equal to that sacrificed; (b) no other way of salvaging it here and now; (c) its protection here and now will not undermine it in the long run. According to him, an action becomes disproportionate if a lesser value is preferred to a higher one, or if in the protection of a greater good an evil is unnecessarily pursued, or if, in certain circumstances, the way of protecting the good will in the long run undermine it (McCormick, 1977, p. 69).

The principle of proportionality depends on the relationship of our reason both to the end and to the means. There must not be any intrinsic contradiction between the basic or ontic good which we want to preserve and the means which we use for that end. As Knauer says, this postulate of non-contradiction between the means and end is a central norm for determining the proportionate reason of any human act (Knauer, 1965, p. 72). That is, there must be ratio between the end and the means. Hence the means receives its moral qualification from its relation to the end (May, 1984, p. 571).

In conclusion, the concept of ordinary and extraordinary means – later evolving in the concept of proportionate and disproportionate means – is a moral concept in the field of theology. It is not a technological concept. It is derived from the day to day problems arising from the field of medicine. The principle behind this concept is an attempt to make a morally good judgment, considering the patient adequately, safeguarding the dignity of human person, at the same time accepting the human limitations and finitude. Such a consideration and acceptance argues for a moral choice among medical procedures related to the patient in his or her existential situation.

C. *Killing, Allowing to Die and Euthanasia*

According to the principle of double effect, a morally relevant difference exists between what one intends and what one merely foresees as a non-intended consequence of one's action or omission. While there are some acts wrong in themselves (such as intentionally killing the innocent), it may nevertheless be permissible to allow death to occur as a non-intended side-effect of some other action that is performed for a proportionate reason. For the principle of double effect to be of any moral interest, it must thus rest on the presupposition that one is not, or is at any rate less or differently, responsible for the foreseen but non-intended consequences of an action or an omission.

To distinguish killing from letting die, supporters of the sanctity-of-life view frequently appeal to the distinction between acts and omissions. Causation is a necessary, though not sufficient, condition of an agent's being responsible for an event. The event must be the consequence of something that the agent does. When an event like death is thus a consequence of an action, such as stabbing or shooting, this is always by way of the causal connection between the given event and the action, and it is this causal connection which enables us to identify the agent and to hold him responsible for the death in question. In others words, the death is a consequence of what the agent "does". But can an agent bring about, or cause, an event only by doing something, or also by doing nothing? Common sense and the law recognize that we are causally, and hence prima facie morally and legally, responsible not only for the consequences of our actions but also for the consequences of at least some of our omissions. Supporters of the sanctity of life Principle agree that (some) omissions can be causes.

Philosophically and legally, omissions have been the subject of much recent debate, especially in terms of the distinction between positive and negative euthanasia, or between killing and letting die. What has made the problem obvious is the development of new medical technologies and our greatly increased ability to sustain the lives of the incurably and terminally ill. Death will often not occur until and unless the agent performs a positive action, such as removing an intravenous drip or turning off a respirator. Is such an action an instance of killing or letting die? Another prominent problem that forces a re-examination of the issue is that of children born with severe birth-defects. While even very seriously disabled infants can now be kept alive, it is often not possible to correct or ameliorate the underlying physical or mental handicap. Because of this, readily available life-sustaining treatments are frequently not employed, and the infants are being allowed to die. What is, however, the basis for sometimes accepting allowing to die and always rejecting killing? The natural response to this question is that killing involves the physician in *doing* something that brings about death, whereas allowing to die or refraining from preventing death does not. The distinction between killing and letting die has recently been overshadowed by the discussion on Euthanasia, essentially as being illustrated in legal developments in the Netherlands. The Court of Law in Alkmaar (The Netherlands) observed in its ruling on euthanasia on May 10, 1983 that: "in ever wider circles, the right of self-determination with regard to the termination of life is becoming accepted". In its official report on euthanasia, the Dutch State Commission also based its judgement on the right of self-determination in offering the opinion that euthanasia may be practiced with impunity if certain procedural rules are satisfied. The debate has indeed developed a growing consensus by focusing on competent patients who request that a doctor either assist them to take their own life (assisted suicide) or actively end life for them (euthanasia). The definition of euthanasia widely accepted in the Netherlands is "the active termination of a patient's life at his or her request, by a physician". Although euthanasia is technically illegal, physicians who adhere to three important conditions recognized by the courts and endorsed by the State Commission on Euthanasia in 1985 are in practice not subject to criminal sanctions: voluntariness, unbearable suffering and consultation (de Wachter, 1989, pp. 3316–3319).

II. SANCTITY AND QUALITY OF LIFE

The moral approach of the majority of Christian ethicists is a critique of two positions, which occupy opposite ends of a spectrum, and which both tend to be reductionistic: "Sanctity of life" (vitalism) and "Quality of Life" (permits cessation of treatment, and even active killing, for the most trivial and even hedonistic of motives). The Roman Catholic tradition has recognized both the sanctity of life – life is indeed sacred – *and* the ethical import of at least some degrees of quality of life – life need not be prolonged under all circumstances, that is, at some point a lack of the ability to carry out the humanly meaningful purposes of life, which some would call a lack of quality of life, means that life can be let go. A personalist approach will first develop a clear understanding of the human person, adequately considered in all his dimensions and relationships. The next step is to develop a medical translation of the "good" being described in anthropological terms.

A. *The Promotion of the Humanly Desireable*

No matter how much our anthropological acknowledgement has enlightened our understanding of the basic structure of the person, no philosophy can even completely reveal the mystery of human personhood. A complementary view of person, must always contain the acknowledgement of the uniqueness and individuality of the person along with necessity of their remaining open to profound inter-personal encounters and also a call to an integration into social relationships that are oriented to working in solidarity for a more dignified society. Each in themselves is part of an ethical program. The unique person along with their interpersonal and social relationships are the three values on which all other values are based. This means that something is of real value only because it serves either the promotion of the person, their inter-personal growth or their solidarity within the community (Schotsmans, 1992). The best medical translation has in our view been presented by James J. Walter.

B. *The Criterion of Purposefulness*

James Walter is convinced that making quality of life judgments is fully consistent with the substance of the longstanding Catholic tradition on the distinction between the ordinary and extraordinary means of preserving life (Walter, 1990, pp. 78–88). Some preliminary distinctions are needed: statements about the quality or qualities of life can be descriptive, evaluative or morally normative. Though descriptive and evaluative statements definitely come to bear on clinical and ethical decision-making, all would agree that the key issue concerns the meaning and validity of morally normative claims about the quality of life.

What the advancements in technology in medicine have done is to call into question the traditional goals of medicine. No doubt, medicine rightfully seeks to prevent death (especially an untimely death), to alleviate pain and physical suffering, and to promote health. Though these are important goals in the application of medical knowledge and skill,

Walter would argue that they must be viewed as subordinate to the more encompassing goal of serving the purposefulness of personal existence. Physicians promote health, prevent death, perform surgery, relieve pain, etc. on behalf of others in order that we as patients might continue in some fashion to pursue values (material, moral and spiritual) that transcend physical life. Pain, disease, general ill health and death either frustrate our desires to pursue these values or make it impossible to pursue them at all. As a consequence, when medicine can no longer promote this goal for a patient at all, or when, by its interventions, medicine will place a patient in a condtion that makes the pursuit of purposefulness too burdensome, then medicine has reached its limits on the basis of its own principal reason for existence. The word "quality" does not and should not primarily refer to a property or attribute of life. Rather, the quality that is at issue is the quality of the relationship which exists between the medical condition of the patient, on the one hand, and the patient's ability to pursue life's goals and purposes understood as the values that transcend physical life, on the other. If we understand the phrase to refer to the quality of a relation and not to a quality of life itself, then the evaluative status of physical life is no longer a central issue in the debate. One of the items that most often remains on a hidden agenda behind this debate, concerns – following Walter – the goals and limits of medicine. Walter proposed already that the central and overarching end of medicine is *to promote and enhance the purposefulness of physical and personal life*. Such a proposal about the goal of medicine does not at all address the worth or value of physical or personal life. What the proposal does do is to state the *raison d'être* of medical interventions and its limits, on the hand, and to give us some insight into what is meant in a general way by the terms "benefits", "burdens", and "best interests" of a patient, on the other hand. Walter is convinced that quality of life judgments should not be construed as judgments about the value of either physical or personal. They are not concerned with assessing qualities or properties that, when present, make life itself valuable. Rather, these judgments are evaluative and normative claims or assessments about *the relation* between the patient's medical condition and the patient's ability to pursue material, moral and spiritual values which transcend physical life but do not give that life its very meaning and worth. As such, they specify concretely the meaning of the terms "benefit", "burdens" and "best interests" of a patient as well as the limits of medical interventions within a given historical and cultural situation.

Whereas all physical life is of equal ontic worth and all personal life is of equal moral value, the quality of the relation between these lives and the pursuit of values is not equal. Due to multiple factors, a number of which have to do with genetic endowment and the ways in which we live our lives and a number of which are dependent on the nurturing and accessibility of values in culture, some people are fortunate enough to attain a high quality of life. Other individuals, regrettably, are not as fortunate, and they must live most of their lives pursuing life's purposes at a less than optimal level. But some have no discernible or such a minimal qualitative relation between their medical condition and the pursuit of values that a growing number of theologians have argued that those in this last category have no moral obligation to prolong their physical lives and thus that all treatment, including artificial nutrition and hydration, can be withdrawn from them. In the past, most, if not all, these lives in such conditions would have been mercifully ended by the underlying pathology, but the intervention of modern medical technology today has not been as merciful.

Two things are important for Walter: first, none of the theologians proposing this view have accepted the active killing of patients. Second, this view does not fall into either a eudaimonistic or an exclusionary use of quality of life judgments. These theologians have not drawn a line at the upper limits of pursuing life's purposes (eudaimonistic use), but they have sought to establish the lowest possible limits of what reasonable people would judge bearable and acceptable vis-à-vis the qualitative relation under consideration (cfr. McCormick). Furthermore, this position does not exclude those who fall below these limits from our ordinary moral obligation not to kill them or to care for them.

The duties and their limits to prolong life that bear on health-care professionals are correlative to the patient's obligations and their limits. When it is determined that a patient no longer has an obligation to prolong physical or personal life, e.g., when the pursuit of values is too burdensome due to an underlying fatal pathology that cannot be removed, then medical personnel do not exclude the patient from their obligations to offer treatment but acquiesce to the limits of the patient's obligations. Thus, other things being equal, when medicine can intervene to ameliorate the quality of the relation between the patient's condition and the pursuit of life's goals, then such an intervention can be considered a benefit to the patient and is in his/her best interests. However, because of the condition of the patient, when a proposed intervention cannot offer the patient any reasonable hope of pursuing life's purposes at all or can only offer the patient a condition where the pursuit of life's purposes will be filled with profound frustration or with utter neglect of these purposes because of the energy needed merely to sustain physical life, then medical intervention can only offer burden and is contrary to the best interests of the patient.

What should be obvious is that quality of life judgments are concerned with what the Vatican Declaration on Euthanasia called the assessment of a due proportion between benefits and burdens (see, Appendix B). However, the proportionality referred to is not about the benefits and burdens of the treatment considered in themselves apart from the patient; rather, the assessment is concerned with the proportionality of benefits/burdens (considered teleologically) that will affect the quality of the relation between the patient's medical condition and his/her pursuit of values. By adopting this view of what quality of life judgments are concerned with, it seems that the so-called two ethics of life are not two but really one.

III. ADMISSION TO AND REMOVAL FROM INTENSIVE CARE

In a personalist view ethical analysis is more than a matter of procedural analysis. It is a question of what is the right thing to do. The exclusive affirmation of the autonomy principle can, indeed, prevent us from seeing the wonder of being human in its multi-sidedness. And this is the fundamental good: the promotion of the patient in all his dimensions and relationships, adequately considered.

It is at the same time the background against which we would like to formulate an ethical stand for what criteria should determine whether a patient is to be admitted to or removed from the benefits of intensive care. The decision on the ethical justifiability of using all possible means to prolong life involves not simply an evaluation of the physical existence but that of the whole person. Therefore, it is good to be aware of an entire complex of values and norms with regard to life. Biological life implies a call to humanization. Our relational-personalistic approach has now to be translated in an ethical guideline for dealing with the problem of excessive medical technology.

First, at a given moment, a situation may arise in which it may be possible to prolong life or the dying process for a shorter or longer period of time by applying medical means or by fighting possible complications. Usually, the ethical question is formulated as follows: on the basis of what argument is it ethically justifiable to refrain from employing life-sustaining measures? This question follows from a certain ethical vision that physicians are bound to do whatever they can. This opinion was reasonable at a time when there were relatively few techniques available for postponing an inevitable death for a certain period of time. Today, however, the possibilities for prolonging biological life have expanded tremendously. It therefore appears necessary to reformulate the ethical question and to ask: on what grounds should a physician artificially prolong an inevitably terminal process? This question presupposes a different ethical background, namely, the understanding that one is only obliged to use the available medical means for the prolongation of life if the well-being of the patient (individually or socially) is genuinely being served. The point of departure here is the respect for the person and the process of dying as interhuman event.

A second observation would be that, when a decision must be made about admitting a patient to or removing a patient from an intensive care, the doctor has to weigh the various values and disvalues which are at stake, including all the important elements: personal and familial as well as medical, and that he/she must consider the extent to which a particular treatment might promote human fulfillment. This proportionate judgment needs to be made in each particular case: one will evaluate the possibilities by considering the type of treatments available, their degree of complexity, the costs, and their possibilities of application along with the result that can be expected, taking into account the condition of the sick person and his or her bodily and moral strength. The attending physician, therefore, will need to consider to what degree any treatment can still be meaningful, guided by a deep respect for facilitating a process of dying that is worthy of human beings. Obviously, one cannot do this alone. The collaboration with colleagues, nurses, patients and family is essential, which is another symbol of responsible involvement.

These are the two fundamental steps in the ethical decision making: assessing whether or not one should sustain life and assessing biological life in view of other values. Then comes the third step: making the decision. The responsibility for this decision lies in our view in the hands of the physician. He/she is indeed the only one in the situation who has the skill, the experience and the responsibility to bear the load of such decisions. Some people would call this an example of medical paternalism. We don't think it is right to react in that way. Parents of dying children are not prepared to make such decisions and to be confronted with this cruel dilemmas. Children of dying patients sometimes also avoid this kind of involvement. Of course, it has been made clear enough that the structure of this decision must be fully relational. The excercise of medicine in such cases, is in our view, indeed, fully relational: as a caring and curing profession medicine is a symbol of the relational structure of the human person, as we have described this.

IV. THE ROMAN CATHOLIC VIEW ON JUSTICE

Pope John XXIII said "that every man has the right to life, to bodily integrity, and to the means which are necessary and suitable for proper development of life; these are primarily food, clothing, shelter, rest, medical care, and finally the necessary social services (Pacem in Terris, # 11)". Human beings recognize that certain basic goods or values are worthy of their allegiance and that the pursuit of these goods constitutes human flourishing. We also recognize that these goods – life and health, friendship and peace, truth and beauty – are the object not only of individual pursuit but are the goals of our common activity. The obligation placed on us by these goods is not only individual but also common. The community of men – like individuals – ought to pursue and respect these values. The community respects these values in much the same way as individuals do. A community should not act against these values as they are realized in the lives of its members and other people and communities, nor should it allow individuals within the community to do so.

When it comes to the pursuing of these values by the community, however, things become more complex. Not all communities need pursue all that is good, and the pursuit of different values in a complex community is undertaken by dividing labor, separating tasks, and thus creating roles. Thus, not everyone in a community pursues the values of the community in the same way. Moreover, this pursuit of values which are common to the entire community by diversification of roles generates positive duties on the part of those who fulfill the various roles and entitlements on the part of others in the community. These roles are founded upon the common commitment of the community to the values which define the roles. This fact implies that the person who carries out a role does so on behalf of the community and for the good of its members insofar as they share in the commitment to the good in question. Thus, the person fulfilling a role has certain duties and members of the community have certain claims upon the benefits achieved through his specialized activities.

This sketch is sufficient to show why there is a right to a degree of health care. The aim of health care is health, which is a perfection of bodily life. As such, it is a component of human life – a basic good for man. Moreover, we live in a complex society which is committed to this value. The medical professions carry out this commitment in a specialized way and should do so for the common good. Society commits resources to achieve this value. Thus, there is an obligation that the medical professions provide the results of their specialized activities to the members of the community. Similarly, the members of the community are entitled to a fair measure of these results: they have a right to health care, and especially to that care which – because they do not as individuals have the expertise of the medical professions – they cannot provide for themselves.

On this account, the right to health care is not a tendentious slogan or a mysterious ethical absolute; rather, it arises from the same source as individual obligations – the human good. The basis for a fair distribution of health care resources is medical need – or, more precisely, that level of need compatible with the fulfillment of the similar needs of other persons in the community. There are other possible bases for a fair distribution, but in the case of health care, need is the most reasonable. Other bases – such as merit or wealth – are only accidentally related to the community's goal of helping its members to promote their health.

To sum up : there is a right to health care. This right does not by itself require that the community commit more of its resources to the health sector. Moreover, it does not demand that everything that some people regard as medical care be provided. Only that care which is directed to health is a matter of entitlement. To demand more is to demand what might not be part of the common good of the community and what might even be opposed to this good, or what might render impossible a common agreement about what constitutes this good. Nor does this right absolve individuals from responsibility for their own health: on the contrary, it presupposes this responsibility. What the right to health care does require is that the members of community committed to heal be provided on an equal basis with the medical care they need. Justice does require that some community response be made (J.M. Boyle Jr., 1987, pp. 643–649). The issue of scarce resources and the quality of health care concerns the problem of allocating, or rationing, scarce medical resources in a just and effective way. It is an excruciatingly complex question. The Roman Catholic tradition, which was virtually the only source of medical ethics up until the 1960's, scarcely mentioned the question. Over the past thirty years or so, the governing context for decisions about treatment has shifted, largely due to the growing influence of medical ethics and of the consumer movement, from that of physician paternalism to that of patient autonomy. The emphasis in recent years on informed consent is characteristic of this shift. But no sooner has patient autonomy gained general acceptance than we are confronted with the need for cost containment which threatens, at least to some considerable extent, to take decision-making power out the hands of both physician and patient, and place it in the hands of society as a whole. In the United States eight-tenths of one percent of the entire gross national product is spent in intensive care units. The American example – with large numbers of Americans without health insurance – illustrates a societal system radiçally stressing the importance of individual liberty and at the same time, the total failure of this system; the social security system of some European countries (like Holland, Belgium etc.) is, however, also confronted with growing dilemmas of cost-containment etc.

It is worth noting that one of the burdens which the Catholic tradition mentions in distinctioning ordinary and extraordianary means is the burden of cost (Wildes). It is morally right, says the Catholic tradition, to choose to forgo costly treatments if we have better things to do with the money. It is true that this tradition was developed long before the advent of third party payors. Today the burdens and benefits most usually considered are those for the patient himself or herself. But third party payors are not always involved, and even when they are, some restrictions for cost containment are justifiable. In any case, the Catholic tradition is clearly open to the possibility that a treatment may rightly be forgone in order to help others with the resources saved.

Despite instances of overlap, there is an easy way to distinguish decisions to forgo treatment when they are made on the basis of proportionality from decisions to forgo based on allocation of resources. The former decisions, the ones we make all the time in hospitals, are ones we ought to make even if resources are infinite. The reason why we do not treat pneumonia in a patient with metastatic cancer is that the treatment is not proportionate, and we ought not treat. The reason why we are morally and ethically justified in removing ventilator and nutritional support from a patient in a persistent vegetative state is that this treatment is not in that person's best interests. We would be ethically justified in forgoing the treatment despite any scarcity of resources. The reason why we give access to or remove patients from the benefits of intensive care, is again that this decision has been made in the patient's best interests.

CONCLUSIONS

Our main task remains, to discover as a community of reasonable persons, what criteria should guide for admitting patients to and removing them from the benefits of intensive care units. In the light of the promotion of the human person in all his dimensions (uniqueness, relational structure and solidarity) and in light of the purposefullness of the medical intervention, physicians and patients have to come to a proportionate decision making process. Justice will be reached if these decisions are made possible in an open structure of an insurance system based on solidarity, in which all citizens are required to participate according to their revenue (cfr. H. Delooz). A growing consensus toward the acceptance of some basic parameters has indeed been reached. These conclusions will point out features for such a system.

First, our health care system must have some form of national health insurance, in which all citizens are required to participate according to their revenue. This system must guarantee all of us, regardless of wealth or employment, access to primary care. We should try as much as possible to eliminate the cost-containment which now occurs by rationing against this access to basic care (D.F. Kelly, 1991, p. 152). In addition, and still on the level of primary care, we must allocate more possibilities to educational attempts at prevention of illness.

Second, the same health insurance program should cover most medical procedures which are reasonably likely to return patients to health. This second requirement is consistent with our criterion of proportionality of medical interventions. The criteria of proportionality must, however, be applied as strictly as possible. What would be unjust is to go about our business as usual, rationing in hidden ways against the most defenseless.

REFERENCES

Beauchamp, T.L., Childress, J.F.: 1989, *Principles of Biomedical Ethics. Third Edition*, Oxford University Press, New York.

Boyle, J.M. Jr.: 1987, 'The concept of health and the right to health care', in Lammers, S.E., Verhey, A., *On Moral Medicine. Theological Perspectives in Medical Ethics*, W.B. Eerdmans Publishing Company, Grand Rapids, pp. 643–649.

de Wachter, M.A.M.: 1989, 'Active euthanasia in the Netherlands', *The Journal of the American Medical Association*, 262, pp. 3316–3319.

Fuchs, J. : 1984, *Christian Ethics in a Secular Arena*, Georgetown University Press, Washington.

Janssens, L.: 1979, 'Ontic evil and moral evil', In : C.E. Curran and R.A. McCormick (eds.), *Readings in Moral Theology no. 1. moral norms and catholic tradition*, Paulist Press, New York, pp. 40–93.

Kelly, D.F.: 1991, *Critical Care Ethics. Treatment Decisions in American Hospitals*, Sheed & Ward, Kansas City.

Kelly, G.: 1950, 'The Duty of Using Artificial Means of Preserving Life', *Theological Studies* 11, 203–220.

Kelly, G.: 1951, 'The Duty to Preserve Life', *Theological Studies* 12, 550–556.

Kelly, G.: 1961, *Medico-Moral Problems*, Catholic Hospital Association, St. Louis.

Knauer, P.: 1965, 'La détermination du bien et du mal moral par le principe du double effect', *Nouvelle Revue Théologique* LXXXVII, pp. 356–376.

Kuhse, H. : 1987, *The Sanctity-of-Life Doctrine in Medicine. A Critique*, Clarendon Press, Oxford.

McCormick, R.A.: 1977, *Ambiguity in Moral Choice*, Marquette University Press, Milwaukee.

May, W.E.: 1984, 'Aquinas and Janssens on the moral meaning of human acts', *The Thomist* 48, 566–606.

Nellikatt, S.: 1991, *Ordinary and Extraordinary Means. An Analysis of the Historical Evolution of the Concept and Its Application to Moral Issues*, Faculty of Theology, Leuven.

Schotsmans, P.: 1992, 'En de mens schiep de mens. Medische (r)evolutie en ethiek', Uitgeverij Pelckmans, Kapellen.

Selling, J.A.: 1986, 'The development of proportionalist thinking', *Chicago Studies*, 25, pp. 165–175.

Selling, J.A.: 1988, *Personalist Morals*, University Press, Leuven.

Walter, J.J., Shannon, T.A. : 1990, *Quality of Life. The New Medical Dilemma*, Paulist Press, New York.

THOMAS R. KOPFENSTEINER

THE NOTION OF CRITERION IN CRITICAL CARE ETHICS

INTRODUCTION

Recent works in the philosophy of science, hermeneutics, and ideology critique have underscored the limitations of the modern conception of rationality. The modern uncoupling of system and lifeworld effected a dichotomy in the understanding of the public and private spheres of life, fact and value, right acting and conceptions of the good life, and reason and faith. In the modern world, theology has little to add to rational and public discourse; any attempt to retrieve a theological voice in medical ethics must be aware of the suspicions of modernity (Verhey and Lammers, 1993; Callahan, 1990).

Moving beyond the polemics of modernity does not mean a reconstruction of premodern or mythic world unity. In what can be called the post modern world, there is no immediate correlation between faith and moral or social practice. This, however, does not lessen the indirect influence that faith can have in argumentative discourse. On the one hand, the Church community must retrieve the normative potential of its religious tradition. On the other hand, the Church community must bring its normative orientations of reality to bear on public discourse about issues of justice, rightness, and goodness (Fiorenza, 1992, p. 79). The scope of this paper is limited to the retrieval of the substantive tradition of the Christian community.

There are two characteristics of Christian morality that provide the context in which to deal with the issue of critical care ethics. First, the moral life of a Christian is not lived by uncritical obedience to a closed system of moral norms. Moral norms embody a more original conception of human flourishing. Moral norms have a prophylactic role; like the walls of a medieval city that guaranteed the protection of its citizens, moral norms protect an underlying vision of human dignity which does not exclude suffering, self-sacrifice, or solidarity with those in need. This anthropological horizon is the hermeneutical key to unlock the meaning of the praxis that the norm sanctions or proscribes. To forget that inner-worldly behavior embodies a more original vision of human flourishing is to work with a truncated notion of the reasoning process.

K. Wm. Wildes (ed.), Critical Choices and Critical Care, 145–161.

Secondly, moral norms cannot be limited to prescriptive language. Moral norms must be embedded in the narrative language of the community (Wadell, 1990). This has a particular relevance when speaking about the notion of criterion in critical care ethics. There are two notions of criterion that can be retrieved from the Christian tradition. On one level, in face of human suffering and the experience of death, normative language must yield to the invitation to enter into solidarity with a brother or sister in need. The compassion and care of the believing community for the sick and dying will be an eloquent testimony to the meaning of Christian living. On a second level, in face of limited resources and the growing number of people who have no access to basic health care the Christian community must retrieve the normative potential of its social teaching. In doing so, the Christian community contributes to the conversation about the allocation of health care resources by offering a critique of the present modality of health care delivery and alternative standards with which to interpret the virtue of justice. A theological approach to critical care will keep both notions of criterion in view. Both will reflect the communitarian character of Christian ethics.

I. THE CONTEXT OF FAITH

The starting point for Christian moral reasoning is reason illuminated by faith (DS 3006). The subjective principle of moral knowledge is *ratio fide illuminata*. By faith is meant the *fides quae* or content of faith. To be sure, what one professes in the Christian creed does not provide specific moral norms to guide inner worldly behavior. Rather, faith acts as a presupposition to reason.

There are what Klaus Demmer calls the "anthropological implications" of faith which determine freedom, guide reason and animate moral action (Demmer, 1987, pp. 17–20). The anthropological implications of faith form the background knowledge of the moral community; they provide the contours of a Christian pattern of perception; they form a habit of mind, and sensitize the believer to perceive the world in a certain way; they inform and elucidate descriptions and interpretations of the world. The anthropological implications of faith form basic human integrative convictions through which experiences are organized. In this context, faith does not replace reason or diminish the need for reflection, but circumscribes the horizon in which moral reason works. The autonomous character of moral reasoning is meant to protect the moral law from being reduced to a heteronomous norm. The autonomy of reason is not understood in a Kantian sense of a presuppositionless reason. Any conception of reason that is sensitive to the exigencies of the post modern world will understand the autonomy of moral reasoning as a relational autonomy (Walter, 1981, pp. 237–77; MacNamara, 1985, pp. 37–66).

A first article of faith is the *imago dei.* Christians profess that each person is created in the image and likeness of God (Gen. 1:26). This is the foundation for the dignity of each person. The dignity owed to each person is not merited; it is not destroyed by sin; it is not diminished by sickness or disability. Moral reasoning probes the practical implications of human dignity. When it is said that medicine and technology are to be used in ways that protect and promote human dignity, this means their use will avoid any sort of arbitrary manipulation or paternalism which would rob individuals of their proper freedom and autonomy in health care decisions.

Secondly, in the creed one professes to believe "in Jesus the Christ, born of the Virgin and made flesh". By becoming like us, Christ has rendered us equal (Gal. 3:28). The Incarnation calls for a new form of human solidarity. It is one of the basic anthropological implications of faith to view all human contact as part of the incarnate presence of God (Demmer, 1987, p. 25). Moral reasoning protects the individual from any form of discrimination. The Incarnation provides a theological premise for drawing close with the sick and dying. In doing so, Christians are animated by the words of Christ, "I was sick and you took care of me" (Mt. 25:36).

A third article of faith that will guide Christian praxis is the belief in the death and resurrection of Jesus Christ. The Christian lives with the confidence that death is not a catastrophe at the end of life. The Christian can embrace death as the culmination of life. By Christ's death and resurrection, a person can be reconciled to death and see that the meaning attributed to death gives significance to life. A Christian's death is anticipated and integrated into daily life. In a real way, Christ's death put death to death. This is the source of Christian hope in face of human suffering and death; suffering and death cannot be eliminated, but they can be given meaning.

The witness of religious communities provides a valuable insight into the meaning of Christian death. The religious take time to gather with a dying sister and help her prepare for an impending death; community members stay with her around the clock, pray with her, and comfort her with Scripture, poetry and songs; the room of the dying becomes like a chapel. The example of the religious teaches us that one should not fear the inevitable confrontation with death; one should fear, rather, being unprepared for death, dying unreconciled with one's life choices, and dying alone and isolated from human love.

From the perspective of the Christian faith, the community's care for the sick is not reduced to the relieving of physical suffering, but is concerned with the integral well being of the person. What is offered to the sick and dying will include human solidarity and compassion (Lk. 10:25–38). The measure of Christian care will not be the technology that is available, but the compassion and love that can be extended to those who experience the fragility and limitations of the human condition in their suffering (John Paul II, 1984). With the growing possibilities of pain management through various therapies, the moral discussion must include interpersonal aspects of human suffering and care. The goal is not to merely relieve suffering, but to humanize it.

These reflections provide the initial guidelines for moral casuistry. It is never the case that death must be staved off at all costs or that life must be prolonged by all technological means available. At the same time, however, one cannot shorten life because it is taken as absurd. Normative reasoning works between these two extremes: human life is never seen as meaningless, but the technology used to maintain life can be morally unnecessary (Thomasma, 1990, pp. 167–95).

Keeping a balance between these two extremes is particularly difficult in an atmosphere of high technology. The technology of critical care medicine is taken for granted in a way that was unimaginable a generation ago. This attitude has the tendency to restrict the moral tradition and its application. In a technological society, the duty to maintain life can be equated with the use of all available technology to maintain life; but to use technology in an uncritical way – that is regardless of the patient's condition – is to make an idol of it and turn the patient into an object. Reflection on the use of technology in critical care medicine cannot deny the limitations of human finitude and mortality.

II. THE TRADITION AND ITS LIMITS

1. The Need to Protect the Patient

The Church's teaching underlines the fundamental right to life of every human person. Life is a fundamental good in that it is the condition of possibility for all other goods. A patient should be certain that the right to life is respected and that no one presumes the authority to abrogate it; the ethos of the medical profession is built around the duty to maintain and protect human life.

Built upon the right to life are others which come into play in medical ethics. The patient, for instance, has the right to truthful information about his or her condition. No doubt the duty to provide this information will depend upon the ability of the patient to receive it. Knowing the truth of one's condition is meant to serve the patient's freedom. While the truth should be told in a compassionate and, at times, gradual way, to withhold the truth from the patient is a subtle form of hegemony of the patient; it is a form of paternalism.

Providing information to the patient is also a special challenge to the family. Faced with a situation in which they have no control, the family members' only response can be one of compassion, love, and solidarity with the sick or dying family member. In fact, the support a family gives to the dying will depend upon the love and support experienced in living. Sadly, it is at the time of sickness and death that people often become aware of the banality and superficiality of their lives together. People realize that life has been lived unauthentically and shallowly. By failing to face the reality of a loved one's imminent death, family members only condemn the dying to further isolation by denying the individual the opportunity to take stock of life, to be forgiven one's transgressions, and to make final farewells (MacIntrye, 1985).

The information to which a patient is entitled includes all reasonable information concerning one's diagnosis, the benefits of the proposed treatment as well as its risks, real or potential harms, possible alternatives, and the consequences of forgoing treatment. This information will be integrated into the values and life goals of the patient; it will be the basis for the informed consent of the patient to accept or forgo treatment. The free and informed consent of the patient is the boundary of the possibilities of medicine. One can recall the axiom: *non salus sed voluntas aegroti suprema lex* (Demmer, 1987, p. 147).

Although the informed decision of the patient binds the physician, this does not mean that the physician is obligated to cooperate with the patient with treatment contrary to the physician's moral or professional standards. There can be a delicate balance between the patient's integral well being that the physician is to protect, and the patient's autonomy which the physician is to respect. The autonomy of the patient is meant to safeguard the patient from burdensome treatment and useless suffering. The will of the patient (*voluntas aegroti*), then, is not one and the same with the health of the patient (*salus aegroti*). The complete identification of the physician's duties with the choice of the patient assumes that the freedom to choose is tantamount to human dignity (Quill, 1991). By accepting this assumption, what goes unheard is the desperate and silent cry for human meaning and solidarity in face of the limits and fragility of the human condition.

2. *The Duty of the Professional*

The professional-patient relationship is characterized by mutual respect, trust and honesty. Moreover, the professional-patient relationship is the context in which medicine is practiced and moral decisions are made. Medicine is a means to pursue life's purposes and goals in so far as those projects involve a healthy and well functioning body (Hauerwas, 1987, p. 522).

Above all, the professional has the duty to protect life (Pius XII, 1957a, p. 146). Though this duty has a strategic preference in conflict situations, it is not the only duty of the professional. The professional's duty to maintain life is balanced with the duty to relieve pain, even when the means to do so will shorten life (Pius XII, 1957a, p. 147; Congregation for the Doctrine of the Faith, 1980). Physical suffering is certainly a part of the human condition, but no one is bound to heroic measures of endurance. No one should suffer a physically painful death. The image of a person in anguishing and unbearable pain needs to be tempered by real and effective means of pain management. In addition, the professional has a responsibility to insure – as much as possible – that the person remains conscious and aware of his or her surroundings (Pius XII, 1957a, p. 145; 1958, p. 694). The patient has the right to embrace death while still conscious; a person should not be deprived of consciousness or the ability to communicate with others merely for the sake of prolonging life. This duty serves the humanization of a person's suffering and protects the person from an uncritical and abusive use of technology (Demmer, 1987, pp. 148–49).

This anthropological framework is the context in which to understand the moral act. The moral act is never reducible to its phenomenal structure, but is interpreted by the anthropological context to which it is bound. Like the words of a language that are understood only in the context in which they are used, moral acts need a background narrative which they not only embody but also constitute. The prophylactic function of the moral norm is transferred to the discussion of the moral act and the determination of the object. For instance, a physician can administer analgesics to alleviate pain even though their administration may hasten death. The object of the act is pain control; it is not mercy killing.

3. Moral Systems and Traditional Distinctions

Because decisions about treatment are moral ones and not merely technical ones, situations will arise that create a doubt about which treatment to provide, withdraw, or withhold. The moral systems can be used in resolving cases of doubt.

Traditionally, moral systems did not resolve theoretical issues. Moral systems artificially simplified a complex situation to provide the agent with the practical certainty needed to act in a responsible way. In their original legalistic context, moral systems are easily equated with a kind of moral minimalism. Among the various moral systems are probabilism and tutiorism (Capone, 1981; Demmer, 1987, pp. 54–59). The moral system of probabilism holds that a solidly probable opinion is a sufficient ground to resolve the doubt in favor of liberty. The guiding rule is *qui probabiliter agit prudenter agit.* On the other hand, is the moral system of tutiorism which holds that in doubt, even when there is an opinion for liberty, one acts responsibly when one follows the safer course. These two moral systems came into play in different situations. The tradition held that probabilism was applicable when there was doubt about the moral character of an action; tutiorism was called into play when there was a fear of fundamental goods being violated. In a personalist context, the retrieval of the moral system of tutiorism will be legitimated in those instances when the fundamental good of life is to be protected.

It is often the case in critical care ethics that life is at stake; the moral system at play is tutiorism. Tutiorism, however, should not be interpreted naively or simplistically. Tutiorism does not mean that life must be protected and maintained at all costs. In retrieving the tradition on death and dying, a condition of its reception that cannot be underestimated is the focus on the sanctity of life within the Church. An exclusive emphasis on the sanctity of life will have the tendency to restrict the categories of the tradition on death and dying; consequently, proponents of the sanctity of life will find a strong ally in technology. Tutiorism comes to mean that life must be maintained at all costs. The Christian tradition on death and dying, however, is distorted when the goal of medical technology is limited to the maintenance of life (Boyle, 1990, p. 68). Admittedly, treatments will be initiated with the hope of success; tutiorism requires it. Yet the *presumptio cedit veritati* and treatment may reach the limits of its purpose. Failure to set limits to the use of technology to preserve life will lead to an indiscriminate use of critical care resources. Tutiorism does not mean rigorism; it does not make the maintenance of human life absolute at the expense of others goods to be considered.

In this broader narrative, when treatment would mean a burdensome state of life, or put the patient in a state where he or she would not recognize family members or familiar settings, the withdrawal or withholding of treatment could be a responsible and justified course of action. This freedom need not be questioned in those cases when a patient is competent to make health care decisions on his or her own, or when an individual has left clear and explicit instructions to this effect either in writing or verbally. When a patient's wishes are not known, the burden of decision falls to family members or loved ones in consultation with professionals. A responsible decision would be based on the values, goals and responsibilities of the patient, or the *sensus communis* in morality.

This is the context in which to introduce the Church's explicit teaching concerning medical treatment. It is well established within the Catholic tradition that life need not be maintained by all medical means or technology available. This conviction stands behind the traditional nomenclature of ordinary/proportionate and extraordinary/disproportionate means of treatment (Pius XII, 1957; Cronin, 1958; Cahill, 1985). Whatever language is preferred, the criterion for determining the means is the integral well being of the patient. The criterion is not the state of medical science or technology alone; if that were the case, what is ordinary today was extraordinary yesterday, and what is extraordinary today will be ordinary tomorrow. Nevertheless, technological advancements in the treatment of a disease obviously diminish the risks involved and offer the hope of greater therapeutic benefits. This means that there is a flexibility in determining the proportionality or disproportionality of the means to maintain life.

A person has a moral obligation to use proportionate means of preserving life; proportionate means are those means that offer a reasonable hope of benefit *and* do not entail an excessive burden to the person, family or others. A person may forgo disproportionate means of preserving life; disproportionate means are those means that do not offer a reasonable hope of benefit *or* entail an excessive burden to the person, family or others (Congregation for the Doctrine of the Faith, 1980; National Conference of Catholic Bishops, 1995). The weighing of the burdens and benefits of treatment will be done against the background of a meaningful life and dignified death. Tutiorism will guarantee that the prognosis of treatments is done with as much certainty as possible; tutiorism does not imply that all means of treatment should be used to maintain life. When disproportionate means of treatment are withheld or withdrawn, the patient is not abandoned. The logic guiding the obligation to humanize suffering shifts from therapy to care. Tutiorism obliges that every effort be taken to make the last stages of life bearable for the person and the family members upon whom the responsibility of care often falls unexpectedly. The moral system of tutiorism carries a tacit challenge to the ecclesial community. The burdens of caring for a dying family member also should be taken into account when evaluating the means of treatment. Each family will face definite limits – financial, moral, psychological – to the care that it can provide. Again, there is an opportunity for the community to humanize suffering in face of the growing technological possibilities to prolong life. The burden that a person or a family can endure depends upon the community support they receive or not. The hospice movement, for instance, is not only an example of care for the dying, but an example of communal support for the families who otherwise would not be able to provide this kind of care on their own. By supporting such a program, the ecclesial community can provide a high standard of care to be imitated. The Christian vision of human dignity not only includes the ability to accept suffering and to transform it by the light of faith, but also the ability to bear the suffering of one another generously and compassionately.

Expense is often one of the possible burdens of treatment. The Catholic tradition recognized the freedom and the responsibility of a person not to invest limited financial resources in a treatment disproportionate to the anticipated results. Again, the community's support is important. It often happens that when a family is not able to pay for a potentially beneficial treatment, public support is mobilized to assist the family in need. It is important to note, however, that though traditional authors listed expense as a criterion by which to determine the burdens of a treatment, it factors in the decision making process in different ways.

A mother may decide to forgo even beneficial treatment if, for instance, it would exhaust a family's financial resources or the money spent could be used for other purposes. Her self-sacrifice can be interpreted as a sign of her commitment or faithfulness to the family. The criterion of expense, however, cannot be transferred univocally to the area of surrogate decision making. For a surrogate decision maker, the expense of treatment is not a sufficient criterion in deciding to withhold or withdraw treatment from another. At present, a surrogate decision maker cannot reject beneficial treatment for another on the basis of cost alone; other burdens must factor in the decision. In a similar way, the tradition's categories of proportionate and disproportionate means will be circumscribed when used to factor the health care costs to a community. The self-sacrifice contemplated by a woman for the sake of her family cannot be imposed on her by society.

Behind the traditional categories of proportionate and disproportionate means, then, there is a clear assumption favoring a limitation to health care resources (Wildes, 1991; 1995). Within the tradition's casuistry a person is not obliged to submit to treatment that offers no benefit or is unreasonably burdensome. In addition, a person has limited access to useless or futile medical treatment. Recent developments in clinimetrics such as APACHE can assess an individual's prognosis in order to aid in the weighing of benefits and burdens, and the determination of futility (Rie, 1995). The question today, however, is whether expensive but potentially beneficial treatment should be withheld through publicly established limits to health care? Can expensive treatment be withheld if medical resources could be used to meet the more basic and urgent health care needs of others in the community? By what criteria can society set limits to the use of technology to preserve life? By itself the traditional nomenclature does not cast a broad enough network of cognitive relationships to answer this question. The tradition's categories and its tacitly assumed limitation to access to health care need to be complemented by other relevant teachings of the Church which can refocus the notion of criterion in order to "enable and ultimately call forth a new set of priorities for resource allocation" (Kilner, 1992, p. 146).

III. THE SOCIAL TEACHING AND HEALTH CARE

1. Human Dignity and the Common Good

Moral reasoning has achieved certain standards behind which a community cannot fall and still hope to humanly fashion itself. This is a context in which to understand John XXIII's statement:

> Every person has the right to life, to bodily integrity, and to the means which are suitable for the proper development of life; these are primarily food, clothing, shelter, rest, medical care, and finally the necessary social services. Therefore, a human being also has the right to security in cases of sickness, inability to work, widowhood, old age, unemployment, or in any other case in which he is deprived of the means of subsistence through no fault of his own (John XXIII, 1963).

When basic health care is denied a person, the dignity of the person is offended. The absence of access to basic health care is destructive of individuals and communities. The American Bishops in their Pastoral Letter, *Health and Health Care* affirm:

> Every person has a basic right to adequate health care. This right flows from the sanctity of human life and dignity that belongs to all human persons, who are made in the image of God. It implies that access to that health care which is necessary and suitable for the proper development and maintenance of life must be provided for all people regardless of economic, social, or legal status (NCCB, 1981).

In this perspective, the provision of health care is a social responsibility and not a matter of voluntary initiative. The resources that are contained in the health care system are not private resources, but public goods that aim to serve the common good. The common good is not a utilitarian notion to describe the good of the majority. The common good is described as "the sum of those conditions of social life which allow social groups and their individual members relatively thorough and ready access to their own fulfillment" (NCCB, 1986, par. 79). The common good is the reason that groups enter into solidarity with others to work on behalf of all. In health care this means that all interested groups – the providers, the insurance companies, the government, the professionals – work to serve in an adequate way the basic needs of all. The challenge is to create the conditions in health care to adequately protect the dignity of every person. Only a commitment to the common good makes a just system of health care possible. To be just, the social organizations in a society must promote the common good. Justice, in other words, does not refer to a principle of non-interference; justice, rather, seeks the common good; justice is realized in the achievement of the common good.

When access to adequate or basic health care is a condition of the common good being realized, a question can be raised as to whether basic health care is a commodity to be controlled exclusively by the marketplace or a social good to which all are entitled. A market orientation, it is argued, would improve efficiency, lower costs, and improve resource distribution through competition. This is countered with the argument that market incentives would distort the professional-patient relationship, generate demands for profitable services, and aggravate the problems of access – especially to the poor who lack the economic and social resources to gain access to the health care needed (McMillan, 1989; Pellegrino, 1988).

2. *The Preferential Option for the Poor*

Another teaching relevant to the allocation of health care resources is the preferential option for the poor. While this language is new to recent social teaching, the notion that the poor are a special concern of God and, therefore, should be a motivating concern for Christians is not new to Christianity. To forget the hungry, the needy, the homeless, those without medical care, and those without hope of a better future is to become like the rich man who pretended not to know the beggar Lazarus lying at his gate [Lk. 16: 19–31] (John Paul II, 1987, par. 42). Conversely, work on behalf of the poor is a sign that the kingdom of God is present on earth.

A first question is who are the beneficiaries of the option for the poor? The understanding of the preferential option for the poor is not limited to mean the "poor in spirit". Reducing poverty to an interior disposition has been used to justify the holding of riches. Yet, poverty also refers to something more than material or economic poverty. "Poverty is not merely the lack of adequate financial resources. It entails a more profound deprivation, a denial of full participation in the economic, social, and political life of society and an inability to influence decisions that affect one's life" (NCCB, 1986, par. 188).

In addition, it must be noted that the preferential option for the poor is not limited to individual behavior. Behind the attempt to limit concern or action on behalf of the poor to the individual morality is the modern separation between the private and public spheres. To the modern mind to show a preference for one group is to disrupt the social order, to undermine the principle of justice, and to serve as a disincentive for competition. Overcoming the separation between the private and public spheres is the first step in clarifying what the preferential option for the poor means. The preferential option for the poor does not mean providing privileges to compensate for the needs of the poor, but entails redressing the structures that cause their need. It is not about giving privileges but removing poverty (Gudorf, 1992, pp. 37–40). Health care offered to the poor in charity is not an example of the preferential option for the poor. The Christian's concern for the poor is to enable the poor to participate more fully in the human community. In this way, the preferential option for the poor is more than temporary assistance; the preferential option for the poor "must be translated at all levels into concrete actions, until it decisively attains a series of necessary reforms" (John Paul II, 1987, par. 43). Offering charity health care to those in need may at times be necessary, but it is not a substitute for providing access to basic health care.

There is another commitment entailed in the preferential option for the poor. The powerlessness that is experienced by the poor is the result of their exclusion from social decision making. The poor are isolated from and have no voice in non-poor society. Consequently, social institutions including health care facilities are experienced by the poor as hostile, threatening, unaccessible. This problem is compounded when health care is controlled by a market oriented delivery system (McMillan, 1989). Community decisions about health care systematically exclude them and often are controlled by interests other than their own. One thinks of how the issue of handicap access to institutions is controlled not only by the desire to recognize the dignity of the handicapped but the costs to the various institutions and businesses. Though to include the poor in social decision making is a tremendous challenge and risk to poor and non-poor alike, it cannot be ignored since the commitment to eliminate poverty and its insidious effects falls upon both.

3. *Social Priorities and the Allocation of Resources*

For the most part, the patient's willingness, the level of medical urgency or need, and an analysis of the reasonable benefits and burdens of a treatment are the criteria for patient selection. Strategies for microallocation decisions also may draw upon pragmatic criteria such as the quantity of resources needed, likelihood of benefit, or the responsibilities of those treated (Kilner, 1992, pp. 230–32).

On the other hand, moral reasoning must protect against discriminatory criteria for the microallocation of resources based solely on age, ability to pay, or a social-value criterion (Callahan, 1987; Kilner, 1990). Patient selection based on these criteria is a contrast experience to moral reasoning based on human dignity and the equality of every person. Choices at the microallocation level, however, are controlled by the macroallocation of resources. The focus changes at this level. This becomes a question about the kind of health care a society wants to pursue and subsidize. The macroallocation of resources reflects society's priorities in allocating resources for health care.

The very question of health care rationing and the specific initiatives that have tried to make rationing public policy should be seen as attempts to solve the larger issue of the inequitable way that health care is financed and delivered (Kelly, 1991, pp. 152–53; Caplan, 1992, pp. 329–32). The issue of health care rationing and initiatives like the Oregon Basic Health Services Act underline the fact that the present modality of health care delivery already rations health care. An ever increasing number of people experience health care rationing because of their inability to pay for care, insurance status, preexisting illness, providers' unwillingness to pay for services even when treatment would prove beneficial, age, gender, race, geographic location, or place of employment (Catholic Health Association, 1991, pp. 7–10; Churchill, 1987, pp. 14–15). The estimates for the uninsured reach to 37 million Americans; in addition to them, there are nearly 65 million underinsured Americans. Public discourse about the macroallocation of resources is an attempt to challenge established priorities for the use of available resources, and recast allocation priorities in a way that does not further burden disadvantaged groups before patient selection criteria come into play (Kilner, 1992, pp. 237–39).

The barrier to health care that is experienced by so many people stands in contrast to the Catholic social tradition which holds that access to basic health care flows from the dignity of the person and serves the common good, and which makes a preferential option for the poor. This religious tradition is not to be interpreted naively; no system of health care delivery can or should satisfy every need with all possible technology. The social tradition of the Church, however, helps erect a cognitive taxonomy that is used to structure the problem. What is sought through these initiatives, then, is a new measure of what constitutes a responsible and tolerable distribution of limited resources.

A first guide is that the responsible stewardship of resources will demand that the available resources be used to meet the needs of the community served. This is best achieved through a participatory process involving people from all levels of society. Secondly, a responsible allocation of health care resources will support a system of delivery that takes into account factors that contribute to ill health. The practice of medicine is incomplete when it is limited to the use of high technology to treat present illnesses; it must include measures that will prevent or limit the onset of illness. A third and related guide that stems from the social teaching of the Church is that a responsible use of health care resources will require the delivery system to regulate or control access to expensive treatments that benefit only a few people or that are of dubious benefit to anyone in order to expand the availability of more comprehensive and basic health care services to all. This is not to reject technological advancements in medicine; those advancements can alleviate the scarcity of resources which makes allocation necessary (Kelly, 1991, p. 161). Rather, it is meant to offer to the discussion on the social expense of critical care technology standards of behavior guided by the social teaching of the Church.

CONCLUSION

The discussion about criteria in critical care ethics should keep in mind two meanings of criterion. On the one hand, is the tradition of means language. Behind this language stands the Christian community's view of life, suffering and death. On the other hand, is the tradition of the common good and the preferential option for the poor. Behind this language stands the community's vision of a just use of health care resources. Since these two levels mutually condition each other, argumentative discourse in critical care ethics must keep both in mind.

The substantive tradition of the believing community, however, is not interpreted in an ahistorical way or in a way that would exempt the community from public discourse. Through public discourse about moral issues, the mutual and reciprocal criticism of both the substantial normative tradition of the believing community and public conceptions of justice takes place.

REFERENCES

Boyle, T.: 1990, 'Intensive care units (ICUs), and ordinary means: Turning virtue into vice', *The Linacre Quarterly* 57, no. 1, 68–77.

Cahill, L.S.: 1985, 'Respecting life and causing death in a medical context', *Concilium*, 179, 29–38.

Callahan, D.: 1987, *Setting Limits: Medical Goals in an Aging Society*, Simon and Schuster, New York.

Callahan: 1990, 'Religion and the secularization of bioethics', *The Hastings Center Report*, Special Supplement: "Theology, Religious Traditions, and Bioethics", 20, no. 4, 2–4.

Caplan, A.: 1992, *If I Were a Rich Man Could I Buy a Pancreas? and Other Essays on the Ethics of Health Care*, Indiana University Press, Bloomington.

Capone, D.: 1981, 'Sistemi morali', in Leandro Rossi and Ambrogio Valsechhi (eds.), *Dizionario enciclopedico di teologia morale*, Paoline, Rome, pp. 1006–1013.

Catholic Health Association: 1991, *With Justice For All?* The *Ethics of Healthcare Rationing*, The Catholic Health Association, St. Louis.

Churchill, L.: 1987, *Rationing Health Care in America: Perceptions and Principles of Justice*, University of Notre Dame Press, Notre Dame.

Congregation for the Doctrine of the Faith: 1980, *Declaration on Euthanasia*, The United States Catholic Conference, Washington, D.C.

Cronin, D.: 1958, *The Moral Law in Regard to Ordinary and Extraordinary Means of Conserving Life*, Pontifical Gregorian University, Rome.

Demmer, K.: 1987, *Leben in Menschenhand: Grundlagen des bioethischen Gesprächs*, Studien zur theologischen Ethik, 23, Herder, Freiburg.

Gudorf, Ch.: 1992, *Victimization: Examining Christian Complicity*, Trinity Press, Philadelphia.

Hauerwas, S.: 1987, 'Authority and the profession of medicine', In: Stephen Lammers and Allen Verhey (eds.), *On Moral Medicine: Theological Perspectives in Medical Ethics*, Eerdmans, Grand Rapids, pp. 522–526.

John XXIII: 1963, Encyclical Letter *Pacem in Terris*, par. 11. Reprinted in *The Gospel of Peace and Justice: Catholic Social Teaching Since Pope John*, presented by Joseph Gremillion, Orbis Books, New York, 1976.

John Paul II: 1984, Apostolic Letter on the Meaning of Human Suffering, *Salvifici Dolores*, The United States Catholic Conference, Washington, D.C.

John Paul II: 1987, Encyclical Letter *Sollicitudo Rei Socialis*, The United States Catholic Conference, Washington, D.C.

Kelly, D.: 1991, *Critical Care Ethics: Treatment Decisions in American Hospitals*, Sheed and Ward, Kansas City.

Kilner, J.: 1990, *Who Lives? Who Dies? Ethical Criteria in Patient Selection*, Yale University Press, New Haven.

Kilner, J.: 1992, *Life on the Line: Ethics, Aging, Ending 'Patients' Lives and Allocating Vital Resources*, Eerdmanns, Grand Rapids.

MacIntyre, A.: 1985, 'The right to die garrulously'. In: Richard Purtill (ed.), *Moral Dilemmas: Readings in Ethics and Social Philosophy*, Wadsworth Publishing, Belmont, pp. 129–134.

MacNamara, V.: 1985, *Faith and Ethics: Recent Roman Catholicism*, Georgetown University Press, Washington, D.C.

McMillan, E.: 1989, 'The health care marketplace: cost-containment at the price of access', *Theology and Public Policy* 1, 18–31.

National Conference of Catholic Bishops: 1981, Pastoral Letter *On Health and Health Care*, The United States Catholic Conference, Washington, D.C.

National Conference of Catholic Bishops: 1986, *Economic Justice for All*, The United States Catholic Conference, Washington, D.C.

National Conference of Catholic Bishops: 1995, *Ethical and Religious Directives for Catholic Health Care Services*, The United States Catholic Conference, Washington, D.C.

Pellegrino, E.: 1988, 'Rationing health care: The ethics of medical gatekeeping'. In: David Thomasma and John Monagle (eds.), *Medical Ethics: A Guide for Health Professionals*, Aspen Publishing, Rockville, MD., pp. 261–269.

Pius XII: 1957a, 'Address to the Ninth Congress of the Italian Anaethesiological Society', *Acta Apostolica Sedis* 49, 129–147.

Pius XII: 1957b, 'Address to doctors and surgeons', *Acta Apostolica Sedis*, 49, 1027–1033.

Pius XII: 1958, 'Address to the international college of neuro-psycho pharmacology', *Acta Apostolica Sedis* 50, 687–696.

Quill, T.: 1991, 'Death and dignity: A case of individualized Decision Making', *The New England Journal of Medicine*, 324, 691–694.

Rie, M.: 1993, 'Ethical issues in intensive care: Criteria for treatment', this volume, pp. 23–56.

Schüssler Fiorenza, F.: 1992, "The Church as a community of interpretation: Political theology between discourse ethics and hermeneutical reconstruction". In: Don Browning and Francis Schüssler Fiorenza (eds.), *Habermas, Modernity, and Public Theology*, Crossroad, New York, pp. 66–91.

Thomasma, D.: 1990, *Human Life in the Balance*, Westminster Press, Louisville.

Verhey, A., Lammers S. (eds.): 1993, *Theological Voices in Medical Ethics*, Eerdmanns, Grand Rapids.

Wadell, P.: 1990, *Friendship and the Moral Life*, University of Notre Dame Press, South Bend.

Walter, J.: 1981, 'The dependence of christian morality on faith', *Église et Théologie*, 12, pp. 237–277.

Wildes, K. Wm.: 1991, 'Health care rationing and insured access: Does catholic tradition have anything to say'? *The Linacre Quarterly* 58, no. 3, 50–58.

Wildes, K. Wm: 1995, 'Conserving life and conserving means: lead us not into temptation', this volume, 105–118.

M. CATHLEEN KAVENY

DISTRIBUTIVE JUSTICE IN THE ERA OF THE BENEFIT PACKAGE: THE DISPUTE OVER THE OREGON BASIC HEALTH SERVICES ACT

INTRODUCTION

Bill Clinton was elected President of the United States at least partially on the basis of his promise to reform the American health care system. As things stand now, no one doubts the need for reform. In a nutshell, the present system is expensive (accounting for about 14% of the Gross National Product); ineffective (the health of U.S. residents does not measure up statistically to the residents of similarly developed nations); and incomplete (over 14% of U.S. citizens lack health insurance).

While there is substantial unanimity about the need for reform, there is no such agreement about the appropriate shape of health care delivery and financing in a reformed health care system. Before the Clinton Health Care Reform bill was unveiled in November of 1993, various groups with urgent and divergent interests in the future of health care in America promoted their own very different proposals for reform. These groups included the American Medical Association, the American Hospital Association, the Catholic Health Association, and the Health Insurance Association of America. Despite their serious differences about the structure of the financing and delivery of health care in a post-reform era, all of these proposals would have provided health care to the needy through some form of basic benefit package. The Clinton Health Plan has provided such a package on a prepaid capitated basis.[1]

From a health care policy perspective, structuring universal access in terms of a "basic benefit package" has at least three advantages. First, we can define the contents of the package in such a way as to encourage the use of cost-effective treatment and discourage expensive and useless procedures.

Second, we can eliminate the arbitrariness and delay which presently characterize the access of the uninsured needy to our health care delivery system. Rather than waiting for a true medical emergency or

K. Wm. Wildes (ed.), Critical Choices and Critical Care, 163–185.

hoping that a hospital would provide necessary services by drawing on its uncompensated care reserves, they would be assured access to the system for a specified set of services.

Third, a basic benefit package approach would allow us to manage in a more responsible manner the funds necessary to provide medical care to the needy. We can take responsibility for our expenditures by coordinating the contents of the benefit package with an up-front allocation of the actuarially determined amount required to provide that package to all who need it.

It is tempting for ethicists to divide the question of distributive justice in the health care arena into two clearly demarcated categories of macroallocation and microallocation. Yet in the coming era of health reform, questions of microallocation and macroallocation will not be easily separated. Whether an individual patient will receive certain treatment or not is likely to depend on whether the basic benefit package (or the practice guidelines and utilization review procedures that guide its application) covers that treatment in his particular case. In turn, the contents of the basic benefit package will likely reflect societal perceptions about the purpose of health care and the best use of health care dollars.

Consequently, before we can understand how to distribute a specific resource such as intensive care in a just manner, we need to understand the moral meaning of distributing health care in the era of the benefit package. To take a few halting first steps toward such understanding will be the goal of this essay. To focus my analysis, I will use the debate surrounding the most detailed and sophisticated effort to develop a basic benefit package to date. It is not found in one of the major national reform proposals, but in the efforts of the state of Oregon to grapple with its own spiraling health care costs.

I. THE OREGON PLAN

Oregonians face the same seemingly intractable health care problems as the rest of the country. As of 1991, the vast majority of the state (2.2 million people) was covered by private health insurance, generally obtained through an employer. 205,000 Oregonians received health insurance under the Medicaid program. But 450,000 Oregonians, or 18% of the population, fell into a gap that left them without coverage.[2]

Oregon's Basic Health Services Act (OBHSA) mounted a three-pronged attack on the health care crisis, which is designed both to control costs and insure universal access to health care by making a basic benefit plan available to all those who lack insurance.[3] Since 65% of the uninsured come from working families, the true keystone of the Oregon Plan is its mandate that all employers make health insurance under the basic benefit plan available to employees and families of employees.[4] The second component of the Oregon Plan establishes a state run and state subsidized high-risk pool which will provide affordable coverage under the basic benefit plan to those who are unable to obtain any type of private health insurance.[5]

But it is the role of the basic benefit plan in the third prong of Oregon's attack on the health care crisis that has generated the most controversy.[6] As it stood before the implementation of the plan, the state's Medicaid program did not come close to covering all persons at or below the poverty level. To be eligible for Medicaid, persons had to qualify under one of numerous and complicated eligibility categories. For example, families with children would be eligible for Medicaid only if their income was at or below 66% of the federal poverty level; childless men and women would not be eligible for Medicaid at all, no matter how poor (Oregon Health Services Commission).

The Oregon Plan proposes to extend Medicaid coverage to all persons below the federal poverty level. It would fund this extension of coverage, at least in part, by trimming the benefits made available to those who qualified for Medicaid under the old system.[7] The minimum benefit package is not identical to the benefits that those enrolled in the pre-reform Medicaid program received. While it provides some additional services, such as dental care, it fails to cover certain procedures which were once available under Medicaid, such as surgery for low-back pain.

The components of the basic benefit package were determined by the Oregon Health Services Commission, whose eleven members were appointed by the governor and began their work in 1989. Defining basic health care as "a floor beneath which no person should fall", the commissioners faced the legislatively mandated task of developing "a list of health services ranked by priority from the most important to the least important".

The Commission began by identifying and prioritizing seventeen general categories of health services. The ranking of these categories was determined by a formula designed to reflect the Commission's collective judgement on two basic questions. Each commissioner was first asked to rate the relative importance of three general attributes of health care services: value to society, value to an individual at risk of needing the service, and the need for that service category in a basic benefit package.[8] Next, the commissioners ranked of the seventeen categories of service in light of each of the three attributes.

Combining the two rankings described above in a modified Delphi technique, a weighted score for each of the seventeen categories was determined for each commissioner. The final ranking for each category was determined by averaging the scores given by the eleven commissioners individually (OHSC, 1991 Appendix 6). The ranking reveals a preference for treatments categories which yield a total cure of acute, probably fatal diseases, such as appendectomy for appendicitis. Treatments which ameliorate without curing chronic diseases received less favorable evaluation.

The next step in the prioritization process was to fill in the seventeen treatment categories with specific disease / treatment pairs. Ultimately, 709 disease / treatment pairs were arranged under the seventeen categories, using a net-benefit of treatment approach to assess their relative importance. The Commission utilized two types of information in determining net-benefit (OHSC, 1991, p. 23).

The first source was data concerning the health outcomes likely to result both with and without treatment for specific health problems. Since the science of outcomes measurement is still in its infancy, the Commission relied on data supplied by professional specialty organizations and individual physicians (OHSC, Appendix D). The second source for the Commission's determination of net benefit was Oregonians' assessment of how great an improvement in quality of life would result from treatment for each condition.[9] In cases where the Commission felt that the statistical ranking of a condition/treatment pair did not reflect its true importance, the Commission rearranged items by hand.

The end result is a list of 709 condition/treatment pairs ranked in order of their priority to the citizens of Oregon. The first item on the list is medical treatment for pneumonia, followed by treatment for tuberculosis, medical and surgical treatment for peritonitis, and removal of a foreign body in the throat. The item in the lowest priority slot is life support for anencephalic babies. Intensive care treatment for premature babies with a birth weight of less than 500 grams at a point earlier than 23 weeks gestation was item 708. It is worth noting that those statistical cutoff points are sharply decisive for coverage purposes in a way that may not reflect the ambiguities of medical reality. Medical care for premature babies weighing 500 grams or more stands very high on the list, ranking in twenty-second place.

Of necessity, the state of Oregon had to draw a similar line in the sand in order to define the contents of the basic benefits package. To meet its budget, the state of Oregon decided to incorporate into the basic benefit package only the first 587 items on the list. Item 587, which made the cut, is medical therapy for esophagitis; item 588, which did not, is thoracic-lumbar laminectomy and medical therapy for intervertebral disk disorders. Medicaid would now cover the former service; persons who needed the latter would be on their own.

II. DISTRIBUTIVE JUSTICE AND THE ETHICAL CONTROVERSY SURROUNDING THE OREGON PLAN

Not unexpectedly, the ethical debate about the justice of Oregon's innovative approach has been substantial and at times heated. Yet it has been unsatisfactory in the respect that its proponents and opponents often seem to be talking past each other. My concern is to explain why this is the case. But first, two examples will suffice to demonstrate this phenomenon. The effect of OBHSA on the poor has generated a large amount of ethical discussion. The opponents of the Act maintain that the very needy women and children who are presently eligible for Medicaid will suffer as a result of its enactment, since they will be denied access to benefits in order to provide coverage under the basic health plan to persons who are slightly better off financially than they are. On the other hand, the proponents of OBHSA claim that the "neediest" in the relevant sense are not in fact those who are covered by Medicaid, but those who do not qualify for Medicaid but nonetheless cannot afford health insurance on their own. Both sides in the disagreement assume that benefiting the most "needy" is at least one of the criterion by which the Oregon plan is to be judged. The core of the dispute is over how to evaluate who is the "most needy".

A second example comes from the hurdles OBHSA has faced under the Americans with Disabilities Act. In August 1992, Secretary of the Department of Health and Human Services Dr. Louis Sullivan refused to grant Oregon the federal waiver required for implementation. He maintained that it discriminated against persons with disabilities on two counts. In both its emphasis on treatment for acute rather than chronic health problems and its preference for treatments that completely cured problems rather than those which simply obviated their effects, Oregon was in effect relegating to a lower priority the needs of the disabled.

Proponents of the Oregon plan responded that the charge of discrimination was unfounded. The disabled were given access to the same benefits as everyone else; in no event would a covered benefit be denied to a person simply because of his or her disability. Furthermore, the quality of well-being scale by which treatments were judged did not give added points for a cure as such, but merely assessed the net benefit of treatment by comparing the patient's pre-treatment and post-treatment states. Here again, the disagreement is not over whether discrimination against the disabled is wrong, but what counts as such. Is it discrimination in the context of health care not to take special account of the needs of the disabled, as well as the modes they have developed of adapting to their handicaps? Or is it morally sufficient to give them the same coverage as everyone else?

III. A MATRIX FOR DISTRIBUTIVE JUSTICE

If progress is to be made toward resolving the disagreements noted at the end of the previous two paragraphs, they need to be placed in a broader context. Most discussions of distributive justice identify the five commonly articulated principles of justice and focus their energies on deciding which is most appropriately applied in the health care arena these principles include the following:

1) to each according to merit
2) to each according to social contribution
3) to each according to how they satisfy the demands of a free market of exchange
4) to each equally
5) to each according to need (Outka, p. 633; Beauchamp & Walters; Walzer, pp. 84–91).

It would be fruitful to discuss in detail the possible interrelations among these five principles; it is my hunch that they may not be as distinct as they seem, depending on one's conception of the relationship between the person and the common good. But I want to focus on a second level of analysis. In order to do so, I will make the assumption that the proper principle of distribution in the context of health care is "to each according to need". As we saw in the two highlights from the debate over the Oregon Plan, to identify the principle of distributive justice as "to each according to need" is merely to begin to define the problem. We still need to define the "needy" and identify what is needed. These questions cannot be answered in isolation, but only in light of a "matrix" of distribution which requires consideration of five interlocking factors:

1) *The Nature of the Persons:* What is the salient feature about "each" of the persons who will receive the good?
2) *The Good Distributed:* What is the salient feature about the good being distributed?
3) *The Time Frame of the Distribution*: What is the time frame over which the distribution will be judged just or unjust?
4) *The Goal of the Distribution*: What is the goal that the distribution is designed to serve?
5) *The Basis of the Distribution*: What is the basis or principle according to which the good will be distributed?

A moment's reflection demonstrates that the concrete patterns of need-based distribution will vary widely depending on the matrix generated in response to the five questions above. To begin with, consider three vastly different distributive situations, all of which fall into the superficially simple category of distributing food according to need in a situation of scarcity, and all of which actually vary greatly because they generate quite different matrices in response to the five questions set forth above.

A. *Lifeboat Ethics*

Suppose a situation which no doubt happens far more often in the seas of philosophy than in real life: there is a shipwreck, after which a number of people are set adrift in a lifeboat with very little food. Suppose that one of the castaways knows that a freighter is scheduled to pass by the spot three weeks from the day of the catastrophe. The castaways are determined to keep everyone alive until help arrives.

Applying the matrix described above yields the following results:

1) *The Nature of the Persons*: The salient feature of the castaways for the purpose of the distribution is not their social status or their educational level; it is their status as biological beings who must eat in order to live.
2) *The Good Distributed*: Similarly, the most important fact about the food being distributed is not that it is artfully arranged or tastes good, but that it sustains life.

3) *The Time Frame of the Distribution*: The time frame over which the distribution is judged just or unjust is the three-week period from the day of the shipwreck to the day the rescuers arrive. It is unlikely that a person who could not eat one day because of illness would forfeit the food he or she would have eaten that day in order to wipe the distributive slate clean and start the calculations all over again the next day. In this context, "needed calories" is determined in reference to the entire three-week period, not with reference to a single day within that period. Correlatively, we would not say that any great injustice was done to the person who was unable to eat one meal at the same time as the rest of the group, so long as the others did not consume his or her allotted calories.
4) *The Goal of the Distribution*: The goal of distribution is extremely focused: to keep the castaways alive until help arrives.
5) *The Basis of the Distribution*: Finally, the basis of the distribution is calories; how much energy each person needs in order to maintain only the most vital biological functions.

B. A Last Cup of Kindness

Switching to another disaster-movie theme, suppose now a slightly different castaway situation. Suppose the group has made its way to a deserted island dominated by an active volcano. Thanks to an extremely learned geologist in their ranks, the castaways they know that twenty-four hours after their landing, they will die in an eruption of the volcano. As with the castaways in the first scenario, these too have a limited amount of food with them. But their just, need-based distribution will take on an entirely different shape from that described above.

1) *The Nature of the Persons*: The most important feature of the castaways in this situation is not their biological makeup but their spiritual and psychological condition. As they confront the certain knowledge of their own imminent deaths, it is entirely irrelevant that they belong to a species which needs food to live. They are beings who search for comfort and meaning in the face of the impending disaster.

2) *The Good Distributed*: The salient feature of the food they bring with them is not its function in to sustaining biological life but its capacity to bring enjoyment and create community and solidarity. In this situation, it is the social meaning of food that is primary.

3) *The Time Frame of the Distribution*: The time frame of the distribution is the twenty-four hours until the volcano erupts. However, because food is primarily a social/emotional good in this scenario, the timing of the distribution within that one-day period is morally salient in a way not true in the first example. More precisely, since the focal meaning of the food here includes a heavy social component, it may be unjust to hold the last communal meal while one person sleeps, even if the others save a slice of his or her favorite pepperoni pizza. In this context, the meaning of the good to be distributed – the food – has become intertwined with the manner and timing of its distribution.

4) *The Goal of the Distribution*: The goal of the distribution is to bring as much peace and comfort to the doomed castaways as possible in their remaining hours. By satisfying their needs (emotional even more than physical) to the extent possible, the aim is to allow the castaways to confront their fate with as much equanimity as they can muster.

5) *The Basis of the Distribution*: The basis of the distribution is not calories, but enjoyment. Despite the difference in caloric value, one castaway will be given pizza and the other jello if these are their respective favorite foods.

C. Building a Life in the Wilderness

The final hypothetical I wish to consider involves a group of pioneers struggling to build a new life in the wilderness, much like the Pilgrims who landed in my own state of Massachusetts. They have a limited supply of food which must last them through the coming long winter.

1) *The Nature of the Persons*: The salient feature about these pioneers is not merely that they need food to maintain basic biological functions. More importantly, survival in their context means an ability to *work* – to build shelter against the winter storms, to mount a defense against a hostile enemy, to begin building the infrastructure of their new society.

2) *The Good Distributed*: Accordingly, the significance of their meager food supply cannot be found in its capacity merely to sustain biological life. Nor, on the other hand, can that significance be found primarily in the social and psychological meaning of food; reality is too harsh and tasks are too pressing for that. Rather, the salient feature of the food is its capacity to nourish the settlers in a broader sense – to give them the energy and strength they need to complete their allotted work. Nutritional balance in the broadest sense is necessary, not simply calories.

3) *The Time Frame of the Distribution*: The time frame of the distribution is more complicated than in the first two examples. On the one hand, the time period over which the scarcity is operative is much longer than in the first two cases: several months rather than weeks or days. On the other hand, the actual duration of the time of scarcity may be so long as to be irrelevant, and "mini-cycles" of distribution like those we face in ordinary life may become the operative time frame. It is doubtful, for example, that the food will be parceled out once at the beginning of the struggle to settle the new land. Decisions about who gets what food will be made on periodic, week-to-week basis as needs and functions shift within the community.

4) *The Goal of the Distribution*: Correlatively, the goal of the distribution is to allow the settlers to contribute to the community upon which their own survival ultimately depends. One might be tempted to charge the distributors with surreptitiously shifting from a need-based to a social-contribution based mode of distribution. But this example shows that these two modes are not always so distinct, in manifesting concretely two important features about the relationship of the individual to the community. First, it is a mistake to divorce entirely the person's identity from his or her place in society. If human beings are essentially social, then one's social roles are not merely accidental attributes, but constitutive feature's of one's identity. Second, individual persons can work together toward a communal goal without the individual being used as a mere means to social ends.[10]

5) *The Basis of the Distribution*: The basis of the distribution is the nutritional needs of the settlers broadly construed. Unlike in the lifeboat situation, two six-foot tall, one hundred eighty-pound men will here receive different amounts and types of food if one is a member of the clergy and the other a lumberjack.

IV. THE DISTRIBUTIVE MATRIX IN A HEALTH CARE CONTEXT

The foregoing discussion was designed to exemplify the fact that even the seemingly simple task of distributing food under the principle "to each according to need" can generate very different results, depending on the concrete matrix of distribution operative in a given case. The problem with distribution in the health care context, of course, is that no one of the three matrices described above is exclusively appropriate.

Like the lifeboat situation, in some cases health care is narrowly focused on maintaining biological life through a short period of crisis. In others cases, such as when someone is old or dying, the meaning of the care can be found in the comfort it provides against loneliness, darkness, weakness and pain. The category of "effectiveness" – at least as it is normally understood – is inapplicable. In many instances, particularly with regard to preventive care, the aim of the distribution is to give persons what they need to fulfill the obligations and roles they have assumed and promise to assume in society.

Precisely because health care, even more than food, is a concept with polyvalent meaning, it is a mistake to think that it can be adequately accounted for in one distributive matrix. It is my belief that much of the intractable quality of the debate between proponents and opponents of the Oregon plan can be accounted for by the fact that both sides are viewing the "point" of health care from the vantage of a different distributive matrix.[11] In this section of the paper, I would like to examine more closely two different matrices of distribution, more or less correlatable to the proponents and opponents of OBHSA.

A. The Nature of the Persons

The proponents of the Oregon plan construe human beings in an essentially Rawlsian fashion. We can also identify as operative in their arguments the moral epistemology that such an anthropology generates. More specifically, we can see the Rawlsian conviction that one's true self is a rational chooser otologically prior to and distinct from the concrete social and familial circumstances in which our lives become enmeshed.[12]

For Rawls, and for many of the proponents of the Oregon plan, it is from such an "original position" that one is to choose the principles of social justice (See Taylor, pp. 159–17 6; Lukes). The basic benefit package is designed to provide minimally decent care, not to any concrete person, but to an abstract Oregonian as rational chooser who does not know his or her age, health status, familial or social position. The survey used to determine how Oregonians felt about various health conditions erased the personality of the subjects of the distribution in at least two respects.

First, the survey process itself created a composite picture of the health care values of the Oregonian citizen that was far removed from any consideration of the context in which those values are made operative. It did not ask why Oregonians believed what they did about treatment preferences, using their responses to this question to fill out and lend context to their survey answers. It merely asked them to rate various health conditions from zero to one hundred.

Second, the form of the questions themselves led to a highly abstract and implausible consideration of specific health problems apart from the concrete context in which they are experienced. For example, one of the questions reads: You can go anywhere and have no limitations on physical or other activity, but have a bad burn over large areas of your body". Another asks "You can go anywhere and have no limitations on physical or other activity, but you have a painful or weak condition of the back or joints". The normal consequence of a bad burn is the limitations on physical and social activity entailed by the pain and disfigurement it causes. It is difficult to imagine what having a bad burn might be like without those limitations. Similarly, it is difficult to conceive how one can have *no* limitations on one's physical activities, yet have a painful or weak back. The difficulties involved in imagining these health problems in a way that is artificially abstracted from their normal consequences is similar to the difficulties entailed when John Rawls asks us to imagine ourselves as rational choosers in an original position that abstracts from the culture, beliefs, and commitments in which those choices take place.

From an epistemological perspective, the Rawlsian nature of the approach appears in the fact that the viewpoints of persons who have a very particular experience in the health care system can be considered not valuable experience but bias. For example, one ethicist writes: "I believe that Oregon was correct in surveying generally healthy citizens, rather than actively ill patients, because the inevitable biases of the latter group would hopelessly confound the rating process (Hadorn, p. 15)".[13]

What tacit assumption supports the belief that ill persons somehow more biased about health care than those who are well? It is easy to view the young and robust as best approximating the Rawlsian original position. Precisely because they are persons who have not fallen prey to the concrete circumstances that makes them subject to one disease rather than another, one can argue that they stand behind the veil of ignorance with regard to their health care needs.

On the other hand, those who worry that the Oregon plan discriminates against the handicapped do not think that one can separate one's disease from oneself in the fashion available to a Rawlsian. Dr. Louis

Sullivan rejected the argument that the Oregon plan merely ranked disease/treatment pairs in a nondiscriminatory, cost-effective manner. Instead, he asserted that the lower ranking for many treatments that would help disabled people reflected a bias against disabled people. Sullivan's position is consistent with an anthropology which is fundamentally opposed to a Rawlsian view. From such a perspective, Gene Outka's suggestion that it is more just for a society faced with limited resources to exclude persons from treatment based on categories of disease becomes far less tenable (Outka, p. 639).

More specifically, if one assumes a non-dualistic anthropology where one's *self* includes one's body, mind, and soul in an inseparably unity, then any serious illness or disability deeply affects one's *self.* A child with Down's Syndrome does not simply "have" a "disability", that "disability" permeates her personality and marks each and every cell of her body. Albeit to a lesser degree, the same can be said regarding persons stricken by cancer or subject to the inexorable process of aging. Their "illness" or "frailty" marks their countenances, stamps their minds, and consumes and channels their energies. They have no "I" which stands immune from the ravages of their afflictions.

Moreover, an anthropology which not only sees human beings as psychosomatic unities, but also as integrally affected by the communities in which they participate, has an additional reason to reject the separation of illness and selfhood. Persons take their identity in large measure from the roles they have assumed in their society; someone *is* a father, lawyer, convict, friend, or killer. It is often the case that serious disabilities in fact mark if not define our social role; think of leprosy in biblical times and AIDS in our own. If this is so, it is impossible to say to such a person: "It's nothing personal, but we've decided as a society not to treat your disease".[14]

Finally, an anthropology which refuses to define selfhood apart from the social and physical circumstances in which it is lived out would have a different view of the experience of the disabled. Instead of viewing it as a source of "bias" toward the treatment of a given disease, it would be construed as an indispensable source of knowledge.

One way viewing experience as a source of knowledge might have affected the configuration of the plan would be in pointing toward a different estimation of the severity of a given disease or symptom. The Oregon Health Commission acknowledged that first-hand experience was a significant factor in the assessment of several symptoms, including trouble learning, trouble breathing, prescribed medication or diet, wearing glasses, and trouble with sexual performance. The Commission noted that those who had experienced each of these problems deemed them not as severe as those who were contemplating them from the outside. Finally, personal experience might not only be relevant to ranking the severity of symptoms or conditions, but also in evaluating the degree of improvement effected by a given treatment.

B. The Good Distributed

What is the salient description of the good being distributed? For those opposed to the Oregon plan, the good is health care itself: particular treatments given to particular patients who need them. Pointing to the case of Coby Howard, a seven-year-old boy who died in 1987 because the state legislature decided to fund prenatal care rather than liver transplants, Courtney Campbell observes: "All it takes is the death of one visible person ... for a decision about health care priorities to appear cold-hearted, capricious and arbitrary, and discriminatory" (Campbell, S: 554). A judgment that the decision was discriminatory entails a view that the "good" being discriminated is actual treatment: Coby did not get the treatment he needed, while some other child with some other disease did.

By contrast, a close examination of the argument of the proponents of the Oregon plan (as well as most other participants in the health reform debate) reveals that the item to be parcelled out is most often described as "a benefit package". On such a view, Coby Howard was not treated discriminatorily since he got the same benefit package as everyone else. Unfortunately but not unjustly, his illness was not covered by the package.

One can see this approach centering on the distribution of a benefit plan operative in the writings of the architect of the Oregon plan, Dr. John Kitzhaber: By guaranteeing that virtually everyone in the state will have access to the health care system, this approach significantly shifted the debate from *who* is covered to *what* is covered, and created a framework in which we could begin to evaluate the effectiveness and appropriateness of the actual services we are buying with our health care dollars".

What does it mean to distribute a "benefit package"? After all, it is an entity of an odd sort. It is not health care itself, but the conditional right to health care, provided that the health care one requires qualifies as a "covered service" under the package (and, of course, that one pays one's premiums). If rational, self-interested persons in the original position were bargaining about the shape of health care in the real world, their conversation would focus on the contents of the benefit package – what care they would receive under what conditions. Thus the notion of distributing a benefit plan and the presupposition of a Rawlsian "original position" as the optimal context of distribution are mutually supportive.[15]

For example, Clark Havighurst has argued that we should view an insurance contract as an instrument of prospective self-denial by rational persons pooling resources to meet future health care needs in a way which will maximize the use of those resources. He writes:

> Allowing for exceptional cases, these [benefit package] protocols would omit – that is exclude from coverage – any specific services whose expected benefits under particular circumstances did not measure up in a rough comparison with the benefits achievable by putting the same funds to alternative medical uses (Havighurst).

On the other hand, to those who claimed that the contents of the basic benefit package would be justified by a societal agreement made in conditions approximating the original position, the opponents of the Oregon plan could point to the truly hypothetical nature of that position with respect to many health care decisions. It is true that healthy persons might be able to approximate its conditions of just distribution with regard to illnesses and injuries to which they still had some chance of succumbing.

But an opponent of the Oregon plan might reasonably fear, for example, that those defining the benefit package would not be able to adopt the stance of the original position with regard to diseases and conditions affecting neonates. First, those actually making decisions would have no self-interested reason for including care for seriously ill neonates within the scope of the basic benefit package, since they had already negotiated that stage of their lives successfully.

Second, they would have positive reason to exclude extremely ill neonates from coverage. As the practice of major insurance companies indicates, persons pooling their funds to provide each with a common benefit package have every reason to refuse to grant new coverage to persons or conditions that they know in advance will cost more money than they contribute. From the position of pure rational self-interest, well-child care insures that young citizens will be paying into the health care pool for years to come; extraordinary means for a desperately sick neonate merely perpetuates the life of someone who is destined not only to never contribute to the pool, but also to draw heavily on its resources. I believe it is the specter of this logic which worries persons such as Dr. Sullivan.

C. The Time Frame of the Distribution

For the proponents of the Oregon Plan, the time frame over which the distribution is to be judged just or unjust can take two forms. Despite seeming diametrically opposed, they are the mirror image of each other. On the one hand, if one sees oneself as distributing a benefit package as such, then the time frame of distribution is simply the moment at which the insurance contract is signed.

On the other hand, if forced to focus on the contents of the package, a proponent of the Oregon approach would insist that the appropriate framework to judge the distribution must be the span of a human life as a whole.

Not everyone is going to benefit equally from the package at every stage of life. Children might benefit a great deal from covered programs for well child care. Those in their twenties and thirties are not likely to need the package at all. The elderly may get some but not all of what would be medically beneficial to them. To a proponent of the Oregon benefit plan approach, the important point for distributive justice is that in the end, *after* everyone has moved through all stages of life, they will have had precisely the same opportunity to receive precisely the same benefits as everyone else.

To the opponents of the Oregon approach, on the other hand, the time frame of the distribution is the time frame of each person's concrete health care needs. It would not matter what health care needs or resources a needy person had consumed in the past – or could be expected to consume in the future. The salient concern would be balancing the claims of those who need care here and now. Accordingly, it would be important that health care be made available to persons at each and every stage of their lives. Moreover, it would be worrisome that the "whole life" view of distribution described in the foregoing paragraph would render it legitimate to exclude one entire stage of life – for example, the elderly – from access to medical care altogether.

D. The Goal of the Distribution

In *Spheres of Justice*, Michael Walzer argues that goods are to be distributed according to their meaning. "If we understand what it is, what it means to those for whom it is a good, we understand how, by whom, and for what reasons it ought to be distributed. All distributions are just or unjust relative to the social meanings of the goods at stake (Walzer, p. 9)". An essential part of the meaning of a good is what it is *good for*. Consequently, a large part of the disagreement between the proponents and opponents of the Oregon plan revolves around the purpose of health care.

For the proponents of the Oregon plan, the emphasis in the term "health care" is properly on *health*. Kitzhaber writes: "Was the objective to guarantee all Oregonians access to health care? Or was it to keep all Oregonians healthy? We believe it is the latter ..." Consequently, Kitzhaber views it proper to assess health care in terms of health status *outcomes* and to evaluate the distribution of resources in the health care industry in reation to other social conditions which also affect health.

> Health care is but a means to an end – not an end in itself. And thus it is important only to the extent that it furthers the end of maintaining, restoring, or improving health ... We cannot improve the health of our nation if, for example, we continue to spend money only on the medical complications of substance abuse, yet ignore the social conditions which lead to addiction in the first place.

In contrast, my sense is that those who are opposed to the Oregon plan place the primary emphasis in "health care" on *care*. The value of such care sits very uneasily in a scale of outcomes management; the meaning of care lies primarily in the relationship it establishes between human beings, not in what that relationship produces externally. James Burtchaell captured the spirit of this approach in its starkest form in describing the response of physicians in the Netherlands to the policies implemented by the Third Reich:

> They knew their job was not to produce a healthy, working population, nor to eliminate the stunted; it was their profession to heal whom they could, alleviate the affliction of those they could not and stand by all whom they served. They would have agreed ... that death is not the ultimate enemy (though perhaps abandonment is). Their dedication, though, would be not to human life ... but to human beings whose lives we heal if we can, but still serve if we cannot (Burtchaell, p. 508).

Just as Kitzhaber's distributive approach seeks to ally "*health* care" with other ways of augmenting the health of the community, so should the advocates of "health *care*" assess the distribution of this good with other social and institutional forms of human concern for one another, such as respite care, homes for the handicapped, etc. My sense, however, is that their distributive framework is less well-developed than that of the OBHSA's proponents.

E. The Basis of the Distribution

Since all five elements in the distribution matrix are integrally related, much of the analysis of the last element has already been completed in the course of discussing the previous four. Hence we can dispose of this element with dispatch. The proponents of the Oregon approach would distribute a benefit package to each person equally. In contrast, the opponents of the Oregon plan would deliver care to a person as he or she needs it.

V. SUGGESTIONS FOR FUTURE THOUGHT

This paper has attempted to describe in broader context the assumptions surrounding the debate about Oregon's pioneering approach to distributive justice in the situation of scarce health care resources. My conclusion is not that either side in the debate is right or wrong *simpliciter*, but that the debate itself shows that health care is not a simple good which can be distributed according to a single matrix. The basic problem with the Oregon approach is not that it takes into account of health outcomes in formulating its policy. Instead, its flaw is its failure to realize that some aspect of the core meaning of health care – the "corporal and spiritual acts of mercy", if you will – cannot be adequately accounted for in the same cost-benefit analysis.

In the community meetings conducted by Oregon Health Decisions, the value of compassion arose in less than half the discussions. Compassion is defined by Oregon Health Decisions as including "a community concern for life, preserving the integrity of an individual and a family and compassion for the vulnerable such as children and the elderly". In the final ranking of seventeen health categories, services motivated by compassion appeared to have been given a somewhat higher priority. For example, "comfort care" ranked seventh in the list, rather than in the bottom third.

My sense is not that Oregonians see human compassion as a less worthwhile virtue, but that the "values" question was posed to the meeting in a way designed to elicit answers which fit into an instrumentalist framework. The participants were asked to discuss "Why is this health care service important to us"? They were consequently led to focus on the treatment or non-treatment of given diseases as an instrumental means to community-wide ends. The question "What should our provision of health care services say to our fellow citizens about why *they* are important"? might have elicited a different set of responses.

I do not want to state sanguinely that the two matrices of distributive justice in health care which this paper has attempted to explore can be easily meshed. On the other hand, I do not want to deny its possibility, either. To close on an optimistic note, I would like to briefly describe an innovative health care corporation which believes it is possible to combine clinical effectiveness, cost effectiveness and compassion in the case of premature neonates.[16] Paidos Health Care Inc., was formed to provide the technological and nursing support services which allows premature neonates with complex medical needs to be discharged early from neonatal intensive care units.

Several studies have shown that low birth weight babies can be discharged as much as two weeks early (at a weight of between 1600 and 1900 grams) if certain conditions are fulfilled: additional teaching and support of the caregivers, backup care by NICU-trained nurses, and regular communication between the medical staff and the caregivers. Paidos provides the families of fragile newborns with a "tightly woven safety net of acute care services and monitoring" that allows them to meet those conditions.

According to its own studies, the Paidos approach is dramatically cost-effective: Estimating that it can reduce the number of days an infant spends in a neonatal intensive care unit by 20%, Paidos believes that it can save the health care system in the state of Texas $54,246,648 dollars over the cost of a year. Dallas County alone would save about four and a half million dollars.

Yet the fundamental philosophy animating Paidos is not cost effectiveness but the proper shape of human care for one another in the context of a health crisis. The environment of the neonatal intensive care unit's deprives premature neonates of essential human interaction, limits parental contact, and bombards them with an assault of light, sound and activity. Accordingly, Paidos has strived to replace the harshness of the neonatal intensive care unit with a fundamentally more humane home environment that not only nurtures the budding relationship of parent and child, but also opens up toward the wider community. Accordingly, Paidos adopts a model of care that it describes as "centered on individual children being cared for by their families, at home, in their own neighborhoods".

I believe that the crucial test of any health care policy will be its ability to incorporate a distributive matrix which adequately accounts for *all* of the goals of a corporation like Paidos.

NOTES

[1] President Clinton's proposed Health Security Act includes a "comprehensive benefit package", defined generally to include hospital services, emergency services, services of physicians and other health professionals, clinical and preventive services, as well as a number of other services. However, the list of comprehensive benefits in an of itself does not tell us under what conditions persons will be able to receive benefits. A complete picture of the benefit package requires scrutiny of applicable utilization review and quality assurance protocols, as well as practice guidelines. It is these protocols and guidelines that will be used to decide whether a particular person should receive a covered benefit (e.g., a day in the hospital) in a concrete situation. By developing its benefit package in terms of "condition-treatment pairs", the Oregon Plan has made explicit many of the decisions that are made *sub rosa* in the administration of all benefits packages, no matter how "general" the coverage may appear at first glance.

[2] "The Oregon Health Plan", an unpublished report of the Oregon Health Services Commission.

[3] The Plan was approved by the Oregon legislature in August 1993. It will take five years to implement fully.

[4] Senate Bill 935 requires all employers to offer a basic benefit package to their employees by 1994. However, when approving the plan in August 1993, the legislature voted to delay the employer mandate until July 1997 for large and mid-sized businesses and until July 1998 for small businesses. The cost of that package will be shared on a 75% employer – 25% employee basis in the case of the employee's own coverage, and on a 50% employer – 50% employee basis in the case of coverage for the employee's dependents. To ease the transition for small businesses, it includes a phased tax credit and start-up allowance.

[5] Senate Bill 534 establishes a subsidized risk pool for the coverage of persons who are uninsurable or insurable only at prohibitively high rates because of pre-existing medical conditions.

[6] In order to implement OBHSA, Oregon needed the federal Health Care Financing Administration (HCFA) to grant it a waiver from several Medicaid regulations. On August 3, 1992, the Secretary of the Department of Health and Human Services (DHHS), Dr. Louis W. Sullivan, denied the waiver. He contended that it violated the Americans with Disabilities Act, since the process by which services were prioritized seemed to assume that "the life of a person with a disability is less than the value of a life of a person without disability". Spencer Rich, "Oregon Medicaid Rationing Program

Rejected as Biased against Disabled", *The Washington Post*, August 4, 1992. However, on March 18, 1993, Donnan Shalala, DHHS Secretary under the Clinton Administration, granted Oregon the waiver subject to certain conditions.

[7] Senate Bill 27 specifies that income will be the sole test for eligibility under Medicaid and raises the income eligibility threshold to 100 percent of the federal poverty level. It also established a Health Services Commission to compile a list of health services which is ranked in order of priority. SB 27 requires that the legislature determine the benefit package for Medicaid by calculating how many items on the list can be funded for all persons below the federal poverty level. In addition, it requires the use of managed care whenever possible and specified that providers are to be paid at reasonable rates (in contrast to the current practice of negotiating deep discounts for Medicaid patients, which encourages providers to shift costs to those with private insurance.)

[8] The Oregon Health Services Commission describes the prioritization process in detail in its *Prioritization of Health Services: A Report to the Governor and Legislature* (1991). At the Request of the Commission, a separate group, Oregon Health Decisions, held 47 community meetings throughout the state, the purpose of which was to elicit community responses to the question: "Why are certain health care services important to us"? Thirteen values emerged. Listed roughly in the order of frequency with which they were mentioned at the meeting, they are: prevention, quality of life, cost-effectiveness, ability to function, equity, effectiveness of treatment, benefits many, mental health and chemical dependency, personal choice, community compassion, impact on society, length of life, and personal responsibility. The Commission then summarized these values under the three more general attributes of value to society, value to an individual at risk of needing the service, and essential to basic health care, (OHSC, 1991, pp. 21–22).

[9] The Health Services Commission contracted with the Survey Research Center of Oregon State University to conduct a random telephone survey to determine the reactions of Oregonians to various health symptoms and functional impairments. The survey used a modified version of Dr. Robert M. Kaplan's Quality of Well-Being (QWB) Scale. Using a survey sample of 1001 persons, the Center asked questions designed to measure Oregonians' assessment of the effect of six functional impairments affecting mobility, physical activity, and social activity, as well as the effect of twenty-three symptoms on one's quality of well-being. Examples of the functional impairments include being confined to a hospital or nursing home, being confined to a wheelchair or walker under one's own control, and being limited in one's recreational activities. The symptoms that were considered range from loss of consciousness to wearing glasses or contact lenses.

[10] This insight is not foreign to the Roman Catholic tradition. As David Hollenbach writes of the teaching found in the papal encyclicals, "The theory presented in the encyclicals is personalist, not individualist, and it recognizes that persons are essentially social and institution building beings" (p. 97). In a more philosophical approach Michael Sandel writes: "I am likely to experience [the enlisting of my assets and abilities] as a case of being used for others' ends and more as a way of contributing to the purposes of a community I regard as my own" (p. 143).

[11] My sense is that those whose emphasis is placed on prevention have built a strong alliance with those who stress effectiveness of treatment. Both of these frameworks are susceptible to consideration under a cost-benefit analysis. Consequently, the battle lines are between those two groups and those who see health care as a quintessentially human, relational service.

[12] He describes the original position as follows: "First of all, no one knows his place in society, his class position or social status; nor does he know his fortune in the distribution of natural assets and abilities, his intelligence and strength, and the like. Nor, again, does anyone know his conception of the good, the particulars of his rational plan of life, or even the special features of his psychology such as his aversion to risk or liability to optimism or pessimism. More than this, I assume that the parties do not know the particular circumstances of their own society" (p. 137).

[13] The breakdown of the community meetings held to elicit the values of Oregon residents reveals that 93 percent of the participants were white, 69 percent were from health care professions, 67 percent had graduated from college and 69 percent earned an annual income of more than $25,000 (Campbell, S: 555).

[14] I do not mean to suggest that this necessarily is what the Oregon Plan is saying to disabled persons. But I do believe it is what persons like Dr. Sullivan *fear* is being said.

[15] The difference principle reads in its entirety: "Social and economic inequalities are to be arranged so that they are both: (a) to the greatest benefit of the least advantaged, consistent with the just savings principle [which protects the claims of future generations], and (b) attached to offices and positions open to all under conditions of fair equality of opportunity (Rawls, p. 302).

[16] My acquaintance with Paidos arose in the context of Ropes & Gray's role as corporate counsel to Paidos. This description of Paidos is taken from its company brochure.

REFERENCES

Beauchamp, Tom and Walters, LeRoy: 1982, *Contemporary Issues in Bioethics*, Wadsworth Publishing Company, Belmont, CA.

Burtchaell, James: 1987, 'How much should a child cost?' in S. Lammers and A. Verhey (eds.), *On Moral Medicine,* Eerdmans, Grand Rapids, MI.

Campbell, C.: 1991, 'Laboratory or Reform: Setting health priorities in Oregon', *Biolaw* 2:50.

Hadorn, David: 1991, 'The Oregon priority-setting exercise: Quality of life and public policy', *Hastings Center Report* (May–June).

Havighurst, C.: 1992, 'Prospective self-denial: Can consumers contract today to accept health care rationing tomorrow?', *University of Pennsylvania Law Review* 140:5, 1775–1808.

Hollenbach, D.: *Claims in Conflict*, Paulist Press, New York.

Kaplan, Robert and Anderson, John: 1988, 'A general health policy model: Update and applications', *Health Services Research* 23, 2.

Lukes, Steven: 1973, *Individualism*, Basil Blackwell, Oxford.

Oregon Health Services Commission (OHSC): 1991, *Prioritizing Health Services: A Report to the Governor and Legislature,* Salem, Oregon.

Oregon Health Services Commission (OHSC): *The Oregon Health Plan*, Salem, Oregon.

Outka, Gene: 1987, 'Social Justice and equal access to health care', in S. Lammers and A. Verhey (eds.), *On Moral Medicine*, Eerdmans, Grand Rapids, MI.

Rawls, John: 1970, *A Theory of Justice*, Harvard University Press, Cambridge.

Sandel, M.: 1983, *Liberalism and the Limits of Justice*, Cambridge University Press, Cambridge.
Taylor, Charles: 1989, *Sources of the Self,* Harvard University Press, Cambridge.
Walzer, Michael: 1983, *Spheres of Justice*, Basic Books, New York.

PART THREE

APPENDICES

APPENDIX A

POPE PIUS XII

THE PROLONGATION OF LIFE

DR. BRUNO HAID, chief of the anesthesia section at the surgery clinic of the University of Innsbruck, has submitted to Us three questions on medical morals treating the subject know as "resuscitation" [la reanimation].

We are pleased, gentlemen, to grant this request, which shows your great awareness of professional duties, and your will to solve in the light of the principles of the Gospel the delicate problems that confront you.

I. PROBLEMS OF ANESTHESIOLOGY

According to Dr. Haid's statement, modern anesthesiology deals not only with problems of analgesia and anesthesia properly so-called, but also with those of "resuscitation". This is the name given in medicine, and especially in anesthesiology, to the technique which makes possible the remedying of certain occurrences which seriously threaten human life, especially asphyxia, which formerly, when modern anesthetizing equipment was not yet available, would stop the heart-beat and bring about death in a few minutes. The task of the anesthesiologist has therefore extended to acute respiratory difficulties, provoked by strangulation or by open wounds of the chest. The anesthesiologist intervenes to prevent asphyxia resulting form the internal obstruction of breathing passages by the contents of the stomach or by drowning, to remedy total or partial respiratory paralysis in cases of serious tetanus, of poliomyelitis, of poisoning by gas, sedatives, or alcoholic intoxication, or even in cases of paralysis of the central respiratory apparatus caused by serious trauma of the brain.

K. Wm. Wildes (ed.), Critical Choices and Critical Care, 189–196.

II. THE PRACTICE OF "RESUSCITATION"

In the practice of resuscitation and in the treatment of persons who have suffered headwounds, and sometimes in the case of persons who have undergone brain surgery or of those who have suffered trauma of the brain through anoxia and remain in a state of deep unconsciousness, there arise a number of questions that concern medical morality and involve the principles of the philosophy of nature even more than those of analgesia.

It happens at times – as in the aforementioned cases of accidents and illnesses, the treatment of which offers reasonable hope of success – that the anesthesiologist can improve the general condition of patients who suffer from a serious lesion of the brain and whose situation at first might seem desperate. He restores breathing either through manual intervention or with the help of special instruments, clears the breathing passages, and provides for the artificial feeding of the patient.

Thanks to this treatment, and especially through the administration of oxygen by means of artificial respiration, a failing blood circulation picks up again and the appearance of the patient improves, sometimes very quickly, to such an extent that the anesthesiologist himself, or any other doctor who, trusting his experience, would have given up all hope, maintains a slight hope that spontaneous breathing will be restored. The family usually considers this improvement an astonishing result and is grateful to the doctor.

If the lesion of the brain is so serious that the patient will very probable, and even most certainly, not survive, the anesthesiologist is then led to ask himself the distressing question as to the value and meaning of the resuscitation processes. As an immediate measure he will apply artificial respiration by intubation and by aspiration of the respiratory tract; he is then in a safer position and has more time to decide what further must be done. But he can find himself in a delicate position, if the family considers that the efforts he has taken are improper and opposes them. In most cases this situation arises, not at the beginning of resuscitation attempts, but when the patient's condition, after a slight improvement at first, remains stationary and it becomes clear that only automatic artificial respiration is keeping him alive. The question then arises if one must, or if one can, continue the resuscitation process despite the fact that the soul may already have left the body.

The solution to this problem, already difficult in itself, becomes even more difficult when the family – themselves Catholic perhaps – insist that the doctor in charge, especially the anesthesiologist, remove the artificial respiration apparatus in order to allow the patient, who is already virtually dead, to pass away in peace.

III. A FUNDAMENTAL PROBLEM

Out of this situation there arises a question that is fundamental from the point of view of religion and the philosophy of nature. When, according to Christian faith, has death occurred in patients on whom modern methods of resuscitation have been used? Is Extreme Unction valid, at least as long as one can perceive heartbeats, even if the vital functions properly so-called have already disappeared, and if life depends only on the functioning of the artificial-respiration apparatus?

IV. THREE QUESTIONS

The problems that arise in the modern practice of resuscitation can therefore be formulated in three questions:

First, does one have the right, or is one even under the obligation, to use modern artificial-respiration equipment in all cases, even those which, in the doctor's judgment, are completely hopeless?

Second, does one have the right, or is one under obligation, to remove the artificial-respiration apparatus when, after several days, the state of deep unconsciousness does not improve if, when it is removed, blood circulation will stop within a few minutes? What must be done in this case if the family of the patient, who has already received the last sacraments, urges the doctor to remove the apparatus? Is Extreme Unction valid at this time?

Third, must be a patient plunged into unconsciousness through central paralysis, but whose life-that is to say, blood circulation-is maintained through artificial respiration, and in whom there is no improvement after several days, be considered "*de facto*" or even "*de jure*" dead? Must one not wait for blood circulation to stop, in spite of the artificial respiration, before considering him dead?

V. BASIC PRINCIPLES

We shall willingly answer these three questions. But before examining them We would like to set forth the principles that will allow formulation of the answer.

Natural reason and Christian morals say that man (and whosoever is entrusted with the task of taking care of his fellowman) has the right and duty in case of serious illness to take the necessary treatment for the preservation of life and health. This duty that one has toward himself, toward God, toward the human community, and in most cases toward certain determined persons, derives from well ordered charity, from submission to the Creator, from social justice and even from strict justice, as well as from devotion toward one's family.

But normally one is held to use only ordinary means – according to circumstances of persons, places, times and culture-that is to say, means that do not involve any grave burden for oneself or another. A more strict obligation would be too burdensome for most men and would render the attainment of the higher, more important good too difficult. Life, health, all temporal activities are in fact subordinated to spiritual ends. On the other hand, one is not forbidden to take more than the strictly necessary steps to preserve life and health, as long as he does not fail in some more serious duty.

VI. ADMINISTRATION OF THE SACRAMENTS

Where the administration of sacraments to an unconscious man is concerned, the answer is drawn from the doctrine and practice of the Church which, for its part, follows the Lord's will as its rule of action. Sacraments are meant, by virtue of divine institution, for men of this world who are in the course of their earthly life, and, except for baptism itself, presuppose prior baptism of the recipient. He who is not a man, who is not yet a man, or is no longer a man, cannot receive the sacraments. Furthermore, if someone expresses his refusal, the sacraments cannot be administered to him against his will. God compels no one to accept sacramental grace.

When it is not known whether a person fulfills the necessary conditions for valid reception of the sacraments, and effort must be made to solve the doubt. If this effort fails, the sacrament will be conferred under at least a tacit condition (with the phrase "*Si capax est*", "If you are capable",-which is the broadest condition). Sacraments are instituted by Christ for men in order to save their souls. Therefore, in cases of extreme necessity, the Church tries extreme solutions in order to give man sacramental grace and assistance.

VII. THE FACT OF DEATH

The question of the fact of death and that of verifying the fact itself (*"de facto"*) or its legal authenticity (*"de jure"*) have, because of their consequences, even in the field of morals and of religion, an even greater importance. What We have just said about the presupposed essential elements for the valid reception of a sacrament has shown this. But the importance of the question extends also to effects in matters of inheritance, marriage and matrimonial process, benefices (vacancy of a benefice), and to many other questions of private and social life.

It remains for the doctor, and especially the anesthesiologist, to give a clear and precise definition of "death" of a patient who passes away in a state of unconsciousness. Here one can accept the usual concept of complete and final separation of the soul from the body; but in practice one must take into account the lack of precision of the terms "body" and "separation". One can put aside the possibility of a person being buried alive, for removal of the artificial respiration apparatus must necessarily bring about stoppage of blood circulation and therefore death within a few minutes.

In case of insoluble doubt, one can resort to presumptions of law and of fact. In general, it will be necessary to presume that life remains, because there is involved here a fundamental right received from the Creator, and it is necessary to prove with certainty that it has been lost.

We shall now pass to the solution of the particular questions.

VII. A DOCTOR'S RIGHTS AND DUTIES

1. Does the anesthesiologist have the right, or is he bound, in all cases of deep consciousness, even in those that are considered to be completely hopeless in the opinion of the competent doctor, to use modern artificial respiration apparatus, even against the will of the family?

In ordinary cases one will grant that the anesthesiologist has the right to act in this manner, but he is not bound to do so, unless this becomes the only way of fulfilling another certain moral duty.

The rights and duties of the doctor are correlative to those of the patient. The doctor, in fact, has no separate or independent right where the patient is concerned. In general he can take action only if the patient explicitly or implicitly, directly or indirectly, gives him permission. The technique of resuscitation which concerns us here does not contain anything immoral in itself. Therefore the patient, if he were capable of making a personal decision, could lawfully use it and, consequently, give the doctor permission to use it. On the other hand, since these forms of treatment go beyond the ordinary means to which one is bound, it cannot be held that there is an obligation to use them nor, consequently, that one is bound to give the doctor permission to use them.

The rights and duties of the family depend in general upon the presumed will of the unconscious patient if he is of age and "*sui juris.*" Where the proper and independent duty of the family is concerned, they are usually bound only to the use of ordinary means.

Consequently, if it appears that the attempt at resuscitation constitutes in reality such a burden for the family that one cannot in all conscience impose it upon them, they can lawfully insist that the doctor should discontinue these attempts, and the doctor can lawfully comply. There is not involved here a case of direct disposal of the life of the patient, nor of euthanasia in any way: this would never be licit. Even when it causes the arrest of circulation, the interruption of attempts at resuscitation is never more than an indirect attempt to cause the cessation of life, and one must apply in this case the principle of double effect and of "*voluntarium in causa.*"

VIII. EXTREME UNCTION

2. We have, therefore, already answered the second question in essence: "Can the doctor remove the artificial respiration apparatus before the blood circulation has come to a complete stop? Can he do this, at least, when the patient has already received Extreme Unction? Is this Extreme Unction valid when it is administered at the given moment when circulation ceases, or even after"?

We must give an affirmative answer to the first part of this question, as We have already explained. If Extreme Unction has not yet been administered, one must seek to prolong respiration until this has been done. But as far as concerns the validity of Extreme Unction at the moment when blood circulation stops completely or even after this moment, it is impossible to answer "yes" or "no".

If, as in the opinion of doctors, this complete cessation of circulation means a sure separation of the soul from the body, even if particular organs go on functioning, Extreme Unction would certainly not be valid, for the recipient would certainly not be a man anymore. And this is an indispensable condition for the reception of the sacraments.

If, on the other hand, doctors are of the opinion that the separation of the soul form th body is doubtful, and that this doubt cannot be solved, the validity of Extreme Unction is also doubtful. But, applying her usual rules: "The sacraments are for men" and "In case of extreme necessity one tries extreme measures", the Church allows the sacrament be administered conditionally in respect to the sacramental sign.

IX. WHEN IS ONE "DEAD"?

3. "When the blood circulation and the life of a patient who is deeply unconscious because of a central physical paralysis are maintained only through artificial respiration, and no improvement is noted after a few days, at what time does the Catholic Church consider the patient 'dead", or when must he be declared dead according to natural law (questions '*de facto*' and '*de jure*')"?

(Has death already occurred after grave trauma of the brain, which has provoked deep unconsciousness and central breathing paralysis, the fatal consequences of which have nevertheless been retarded by artificial respiration? Or does it occur, according to the present opinion of doctors, only when there is complete arrest of circulation despite prolonged artificial respiration?)

Where the verification of the fact in particular cases is concerned, the answer cannot be deduced from any religious and moral principle and, under this aspect, does not fall within the competence of the Church. Until an answer can be given, the question must remain open. But considerations of a general nature allow us to believe that human life continues for as long as its vital functions – distinguished from the simple life of organs – manifest themselves spontaneously or even with the help of artificial processes. A great number of these cases are the object of insoluble doubt, and must be dealt with accordingly to the presumptions of law and of fact of which We have spoken.

May these explanations guide you and enlighten you when you must solve delicate questions arising in the practice of your profession. As a token of divine favors which We call upon you and all those who are dear to you, We heartily grant you Our Apostolic Blessing.

Reprinted with permission of the editors from *The Pope Speaks*, Vol. 4, 1958.

APPENDIX B

CONGREGATION FOR THE DOCTRINE OF THE FAITH

DECLARATION ON EUTHANASIA

The rights and values pertaining to the human person occupy an important place among the questions discussed today. In this regard, the Second Vatican Ecumenical Council solemnly reaffirmed the lofty dignity of the human person, and in a special way his or her right to life. The council therefore condemned crimes against life. The council therefore condemned crimes against life "such as any type of murder, genocide, abortion, euthanasia, or willful suicide" (*Gaudium et Spes, n.* 27).

More recently, the Sacred Congregation for the Doctrine of the Faith has reminded all the faithful of Catholic teaching on procured abortion.[1] The congregation now considers it opportune to set forth the church's teaching on euthanasia.

It is indeed true that in this sphere of teaching the recent popes have explained the principles and these retain their full force;[2] but the progress of medical science in recent years has brought to the fore new aspects of the question of euthanasia, and these aspects call for further elucidation on the ethical level.

In modern society, in which even the fundamental values of human life are often called into question, cultural chance exercises an influence upon the way of looking at suffering and death; moreover, medicine has increased its capacity to cure and to prolong life in particular circumstances, which sometimes give rise to moral problems.

Thus people living in this situation experience no little anxiety about the meaning of advanced old age and death. They also begin to wonder whether they have the right to obtain for themselves of their fellowmen and " easy death", which would shorten suffering and which seems to them more in harmony with human dignity.

A number of episcopal conferences have raised questions on this subject with the Sacred Congregation for the Doctrine of the Faith. The congregation, having sought the opinion of experts on the various aspects of euthanasia, now wishes to respond to the bishops' questions

K. Wm. Wildes (ed.), Critical Choices and Critical Care, 197–204.

with the present declaration, in order to help them to give correct teaching to the faithful entrusted their care and to offer them elements for reflection that they can present to the civil authorities with regard to this very serious matter.

The considerations set forth in the present document concern in the first place all those who place their faith and hope in Christ, who through his life, death and resurrection has given a new meaning to existence and, especially to the death of the Christian, as St. Paul says: "If we live, we live to the Lord, and if we die, we die to the Lord" (Rom. 14:8 cf. Phil. 1:20).

As for those who profess other religions, many will agree with us that faith in God the Creator, provider and lord of live – if they share this belief – confers a lofty dignity upon every human person and guarantees respect for him or her.

It is hoped that this declaration will meet with the approval of many people of good will, who philosophical or ideological differences notwithstanding, have nevertheless a lively awareness of the rights of the human person. These rights have often in fact been proclaimed in recent years through declarations issued by international congresses;[3] and since it is a question here of fundamental rights inherent in every human person, it is obviously wrong to have recourse to arguments from political pluralism or religious freedom in order to deny the universal value of those rights.

I. THE VALUE OF HUMAN LIFE

Human life is the basis of all goods and is the necessary source and condition of every human activity and of all society. Most people regard life as something sacred and hold that no one may dispose of it at will, but believers see in life something greater, namely a gift of God's love, which they are called upon to preserve and make fruitful. And it is this latter consideration that gives rise to the following consequences:

1. No one can make an attempt on the life of an innocent person without opposing God's love for that person, without violating a fundamental right, and therefore without committing a crime of the utmost gravity.[4]

2. Everyone has the duty to lead his or her life in accordance with God's plan. That life is entrusted to the individual as a good that must

bear fruit already here on earth, but that finds its full perfection only in eternal life.

3. Intentionally causing one's own death, or suicide, is therefore equally as wrong as murder; such an action on the part of a person is to be considered as a rejection of God's sovereignty and loving plan. Furthermore, suicide is also often a refusal of love for self, the denial of the natural instinct to live, a flight from the duties of justice and charity owed to one's neighbor, to various communities or to the whole of society – although, as is generally recognized, at times there are psychological factors present that can diminish responsibility or even completely remove it.

However, one must clearly distinguish suicide from that sacrifice of one's life whereby for a higher cause, such as God's glory, the salvation of souls or the service of one's brethren, a person offers his or her own life or puts it in danger (cf. Jn. 15:14).

II. EUTHANASIA

In order that the question of euthanasia can be properly dealt with, it is first necessary to define the words used.

Etymologically speaking, in ancient times euthanasia meant an easy death without severe suffering. Today one no longer thinks of this original meaning of the word, but rather of some intervention of sickness or of the final agony are reduced, sometimes also with the danger of suppressing life prematurely. Ultimately, the word euthanasia is used in a more particular sense to mean "mercy killing", for the purpose of putting an end to extreme suffering, or saving abnormal babies, the mentally ill or the incurable sick from the prolongation, perhaps for many years, of a miserable life, which could impose too heavy a burden on a their families or on society.

It is therefore necessary to state clearly in what sense the word is used in the present document.

By euthanasia is understood an action or an omission which of itself or by intention causes death, in order that all suffering may in this way be eliminated. Euthanasia's terms of reference, therefore, are to be found in the intention of the will and in the methods used.

It is necessary to state firmly once more that nothing and no one can in any way permit the killing of an innocent human being, whether a fetus or an embryo, an infant or and adult, and old person, or one suffering from an incurable disease, or a person who is dying. Furthermore, no one is permitted to ask for this act of killing, either for himself or herself or for another person entrusted to his or her care, nor can he or she consent to it, either explicitly or implicitly. Nor can any authority legitimately recommend or permit such an action. For it is a question of the violation of the divine law, an offense against the dignity of the human person, a crime against life and an attack on humanity.

It may happen that by reason of prolonged and barely tolerable pain, for deeply personal or other reasons, people may be led to believe that they can legitimately ask for death or obtain it for others. Although in these cases the guilt of the individual may be reduced or completely absent, nevertheless the error of judgment into which the conscience falls, perhaps in good faith, does not change the nature of this act of killing, which will always be in itself something to be rejected.

The pleas of gravely ill people who sometimes ask for death are not to be understood as implying a true desire for euthanasia; in fact it is almost always a case of an anguished plea for help and love. What a sick person needs, besides medical care, is love, the human and supernatural warmth with which the sick person can and ought to be surrounded by all those close to him or her, parents and children, doctors and nurses.

III. THE MEANING OF SUFFERING FOR CHRISTIANS AND THE USE OF PAINKILLERS

Death does not always come in dramatic circumstances after barely tolerable sufferings. Nor do we have to think only of extreme cases. Numerous testimonies which confirm one another lead one to the conclusion that nature itself has made provision to render more bearable at the moment of death separations that would be terribly painful to a person in full health. Hence it is that a prolonged illness, advanced old age, or a state of loneliness or neglect can bring about psychological conditions that facilitate the acceptance of death.

Nevertheless, the fact remains that death, often preceded or accompanied by severe and prolonged suffering, is something which naturally causes people anguish.

Physical suffering is certainly an unavoidable element of the human condition; on the biological level, it constitutes a warning of which no one denies the usefulness; but, since it effects the human psychological makeup, it often exceeds its own biological usefulness and so can become so severe as to cause the desire to remove it at any cost.

According to Christian teaching, however, suffering, especially suffering during the last moments of life, has a special place in God's saving plan; it is in fact a sharing in Christ's passion and a union with the redeeming sacrifice which he offered in obedience to the Father's will. Therefore one must not be surprised if some Christians prefer to moderate their use of painkillers, in order to accept voluntarily at least a part of the sufferings and thus associate themselves in a conscious way with the sufferings of Christ crucified (cf. Mt. 27:34).

Nevertheless it would be imprudent to impose a heroic way of acting a general rule. On the contrary, human and Christian prudence suggest for the majority of sick people the use of medicines capable of alleviating or suppressing pain, even though these may cause as a secondary effect semiconsciousness and reduced lucidity. As for those who are not in a state to express themselves, one can reasonably presume that they wish to take these painkillers, and have them administered according to the doctor's advice.

But the intensive use of painkillers is not without difficulties, because is not without difficulties, because the phenomenon of habituation generally makes it necessary to increase their dosage in order to maintain their efficacy. At this point it is fitting to recall a declaration by Pius XII, which retains its full force. In answer to a group of doctors who had put the question; "Is the suppression of pain and consciousness by the use of narcotics ...permitted by religion and morality to the doctor and the patient (even at the approach of death and if one foresees that the use of narcotics will shorten life)"?

The pope said: "If no other means exist, and if, in the given circumstances, this does not prevent the carrying out of other religious and moral duties: Yes".[5] In this case, of course, death is in no way intended or sought even if the risk of it is simply to relieve pain effectively, using for this purpose painkillers available to medicine.

However, painkillers that cause unconsciousness need special consideration. For a person ont only has to be able to satisfy his or her moral duties and family obligations; hc or shc also has to preparc himself or herself with full consciousness for meeting Christ. Thus Pius XII warns; "It is not right to deprive the dying person of consciousness without a serious reason".[6]

IV. DUE PROPORTION IN THE USE OF REMEDIES

Today it is very important to protect, at the moment of death, both the dignity of the human person and the Christian concept of life against a technological attitude that threatens to become an abuse. Thus some people speak of a "right to die", which is an expression that does not mean the right to procure death either by one's own hand or by means of someone else, as one pleases, but rather the right to die peacefully with human and Christian dignity. From this point of view, the use of therapeutic means can sometimes pose problems.

In numerous cases, the complexity of the situation can be such as to cause doubts about the way ethical principles should be applied. In the final analysis, it pertains to the conscience either of the sick person, or of those qualified to speak in the sick person's name, or of the doctors to decide the light of moral obligations and of the various aspects of the case.

Everyone has the duty to care for his or her own health or to seek such care from others. Those whose task it tis to care for the sick must do so conscientiously and administer the remedies that seem necessary or useful.

However, is it necessary in all circumstances to have recourse to all possible remedies?

In the past moralists replied that one is never obliged to use "extraordinary" means. This reply, which as a principle still holds good, is perhaps less clear today by reason of the imprecision of the term and the rapid progress made in the treatment of sickness. Thus some people prefer to speak of "proportionate" and "disproportionate" means.

In any case, it will be possible to make a correct judgment as to the means by studying the type of treatment to be used, its degree of complexity or risk, it cost and the possibilities of using it, and comparing these elements with the result that can be expected, taking into account the state of the sick person and his or her physical and moral resources.

In order to facilitate the application of these general principles, the following clarification can be added:

– If there are no other sufficient remedies, it is permitted, with the patient's consent, to have recourse to the means provided by the most advanced medical techniques, even if these means are still at the experimental stage and are not without a certain risk. By accepting them, the patient can even show generosity in the service of humanity.

- It is also permitted, with the patient's consent, to interrupt these means where the results fall short of expectations. But for such a decision to be made, account will have to be taken of the reasonable wishes of the patient's family, as also of the advice of the doctors who are specially competent in the matter. The latter may in particular judge that the investment in instruments and personnel is disproportionate to the results foreseen; they may also judge that the patient strain or suffering out of proportion with the benefits which he or she may gain from such techniques.
- It is also permissible to make do with the normal means that medicine can offer. Therefore one cannot impose on anyone the obligation to have recourse to a technique which is already in use but which carries a risk or is burdensome. Such a refusal is not the equivalent of suicide; on the contrary, it should be considered as an acceptance of the ham condition, or a wish to avoid the application of a medical procedure disproportionate to the results that can be expected, or a desire not to impose excessive expense on the family or the community.
- When inevitable death is imminent in spite of the means used, it is permitted in conscience to take the decision to refuse forms of treatment that would only secure a precarious and burdensome prolongation of life, so long as the normal care due the sick person in similar cases is not interrupted. In such circumstances the doctor has no reason to reproach himself with failing to help the person in danger.

CONCLUSION

The norms contained in the present declaration are inspired by a profound desire to serve people in accordance with the plan of the Creator. Life is a gift of God, and on the other hand, death is unavoidable; it is necessary therefore that we, without in any way hastening the hour of death, should be able to accept it with full responsibility and dignity. It is true that death marks the end of our earthly existence, but at the same time it opens the door to immortal life. Therefore all must prepare themselves for this event in the light of human value, and Christians even more so in the light of faith.

As for those who work in the medical profession, they ought to neglect no means of making all their skill available to the sick and the dying; but they should also remember how much more necessary it is to provide them with the comfort of boundless kindness and heartfelt charity. Such service to people is also service to Christ the Lord who said: "As you did it to one of the least of these my brethren, you did it to me: (Mt. 25:40).

At the audience granted to the undersigned prefect, His Holiness Pope John Paul II approved this declaration, adopted at the ordinary meeting of the Sacred Congregation for the Doctrine of the Faith, and ordered its publication.

Rome, the Sacred Congregation for the Doctrine of the Faith, May 5, 1990.

Cardinal Franjo Şeper
Prefect
Archbishop Jerome Hamer, O.P.
Secretary

NOTES

[1] "Declaration Procured Abortion", Nov. 18, 1974: AAS 66 (1974), pp. 730–747.

[2] Pius XII, "Address to those attending the Congress of the International Union of Catholic Women's Leagues, " Sept. 11, 1947: AAS 39 (1947), p. 483; "Address to Midwives", Oct. 29, 1951: AAS 43 (1951), pp. 853–854; "Speech to the Members of the International Office of Military Medicine Documentation", Oct. 19, 1953: AAS 45 (1953), pp. 744–754; "Address to those taking part in the ninth Congress of the Italian Anaesthesiological Society,: Feb. 24, 1957: AAS 49 (1957), p.146; cf. also "Address on Re-animation", Nov. 24, 1957: AAS 49 (1957), pp. 1027–1033; Paul VI, "Address to the members of the U.N. Special Committee on Apartheid", May 22,1974: AAS 66 (1974), p.346; John Paul II: "Address to the Bishops of the United States of America", Oct. 5, 1979: AAS 71 (1979), p. 1225.

[3] One thinks especially of Recommendation 776 (1976) on the rights of the sick and dying, of the Parliamentary Assembly of the council of Europe at its 27th ordinary session; cf. Sipeca, no. 1, March 1977 pp. 14–15.

[4] We leave aside completely the problems of the death penalty and of war, which involve specific considerations that do not concern the present subject.

[5] Pius XII, "Address" Feb. 24, 1957: AAS 49 (1957), p. 147.

[6] Pius XII, *ibid.*, p. 145, cf. "Address", Sept. 9, 1958: AAS 50 (1958), p. 694.

Originally Published
United States Catholic Conference, June 26 1980
Reprinted with Permission

APPENDIX C

SOCIETY OF CRITICAL CARE MEDICINE ETHICS COMMITTEE

CONSENSUS STATEMENT ON THE TRIAGE OF CRITICALLY ILL PATIENTS

It is likely that the demand for health care service will always exceed the supply. In this context of relative scarcity, it is appropriate to develop explicit guidelines to help facilitate the fairest use of these services.[1–11] This statement offers critical care providers principles and guidelines for the distribution of intensive care resources among individual patients.

I. JUSTIFICATION

The United States and other countries face difficult questions regarding access to, delivery of, and payment for health care services, as well as the proportion of health care expenditures that should be appropriated for critical care. These issues are not likely to be resolved easily or quickly nor will these general issues be addressed herein. Regardless of when and how these broader issues are resolved, individual health care organizations and providers will continually face a disparity between demand for and availability of critical care facilities. Demand is created by the inclination of critically ill patients and their families to seek, and their physicians to provide, intensive care. The advancing age of the population is likely to increase the demand for services as more elderly individuals who are frail, chronically ill, and subject to life-threatening illness become potential candidates for critical care. the development of new pharmaceutical products and technological devices often makes the care provided more expensive. At the same time that demand for and expense of care are increasing, the capacity to meet demand is constrained by inadequate reimbursement, restricted growth of health care facilities, and personnel shortages.

In the setting of these constraining conditions, individual institutions and individual providers of critical care must use some moral framework for distributing the resources at hand. Sometimes resources are

K. Wm. Wildes (ed.), Critical Choices and Critical Care, 205–213.

more obviously limited than at other times, as evidenced by the absence of an available bed in the intensive care unit (ICU). The guidelines provided herein are intended to be applicable whether or not an immediate shortage is apparent, because their continuous use will lead to more consistently equitable and efficient critical care. It is recognized that limiting care of critically ill patients during acute shortages is more likely to result in adverse consequences for individual patients than limiting care during times without shortages. While various terms such as "triage", "rationing", or "allocation of resources" have been used to describe the distribution of limited goods and services, we will use the term "triage".[2–16] This term has been chosen because it conveys a well-established process in medicine of finding the most appropriate disposition for a patient based on an assessment of the patient's illness and its urgency. "Triage" derives from the French verb *trier*, meaning to pick, to sort, or to select. The first medical application of the word was in the French military, where *hospital de triage* meant a sorting station for injured soldiers. There are at least six major types of triage, depending on the location of the patient: (1) prehospital, (2) catastrophic, (3) emergency department, (4) intensive care, (5) battlefield, and (6) waiting list for lifesaving treatments in short supply such as dialysis or organ transplantation. We particularly endorse this process when it is based on a sound understanding of the probable outcome of the patient's illness, the availability of therapeutic modalities, the impact of therapy on outcome, and a judgment of the benefits and burdens of the therapy for the patient, the patient's family, and society.

Triage of critically ill patients that may limit individual patient and physician choices is justified when (1) the policy is aimed at achieving benefits for individual patients, the health care institution, and society and (2) the policy is announced in advance to notify the public.

II. PRINCIPLES

General Principles

Guidelines should be formulated for expected categories of triage that may reasonable be anticipated. They should articulate the principles, justifications, and mechanisms pertaining to each situation as explicitly as possible.

Notwithstanding the pressures exerted on health care providers to limit resource consumption during real or perceived conditions of scarcity, physicians should remain staunch advocates of their patient's best interests. Conflicts of interest between the health care provider's role as gatekeeper and patient advocate may occur and should be anticipated. Because conflicts of interest cannot be altogether eliminated, institutions should attempt to separate the roles of gatekeeper and advocate when possible.

Physicians, nurses, and administrative staff should collaborate in framing guidelines and recommendations for the prudent use of scarce resources in their own institutions. The formulation and prolongation of explicit triage policies should help to minimize the necessity for ad hoc triage decisions in individual cases.

Health care providers, patients, policymakers, and the public at large must recognize that various treatments, including life-prolonging treatments, may be denied. Health care insitutions may justifiably restrict the availability of certain services to use limited resources more effectively or to enhance equity in allocating them.[10,17,18]

Priorities for patient selection for triage should be reflected in ICU policies. Priorities should weight the scope of services available, patient inclusion and exclusion criteria,[19] and institutional triage policies. Established standards of critical care should not be compromised.

Critical care units should, in general, be reserved for those patients who have a "reasonable prospect of substantial recover".[20(p799)] An exception may be patients being supported for organ donation. The scope of ICU services should be limited to diagnostic and therapeutic procedures from which a benefit is anticipated.

Each hospital should develop a policy for the accommodation for critically ill patients when all ICU beds capable of being staffed are full. This overflow policy should be prepared in advance so as to identify suitable areas for temporary overflow.

If sufficient ICU beds are not available and appropriate triage does not resolve the shortage, every effort should be made to provide analogous levels of care in other settings. Hospitals should provide alternative solutions for patients in need of life support and/or monitoring who can be reasonably cared for outside the ICU setting.

Basic ICU services of known therapeutic benefit should receive priority over investigational ICU services that are not of proven value.

Specific Principles

The foremost consideration in triage decisions is the expected outcome of the patient in terms of survival and function, which turns on the medical status of the patient. In general, patients with good prognoses for recovery have priority over patients with poor prognoses. While uncertainty of prognosis is a crucial problem in critical care, providers should utilize predictive instruments with a full understanding of their strengths and limitations. Decisions to be made between patients with equivalent prognoses should be made on a first come, first served basis.

Priority for admission to an ICU should correlate with the likelihood that ICU care will benefit the patient substantially more than non-ICU care. Patients with very poor prognoses and little likelihood of benefit should not be admitted. Patients with very good prospects with or without ICU care also should not be admitted.

Benefit and futility have been variously defined by authors based on the goals of therapy and the likelihood of success. They are not specifically defined because they are subjective, value laden terms for which there is no agreement.[21–23] Factors that should be considered in determining benefit and utility for triage decisions include the following:

- likelihood of a successful outcome;
- patient's life expectancy due to disease(s);
- anticipated quality of life of the patient;
- wishes of the patient and/or surrogate;
- burdens for those affected, including financial and psychological costs and missed opportunities to treat other patients;
- health and other needs of the community; and
- individual and institutional moral and religious values.

All triage decisions made for individual patients on grounds of scarcity must be made explicitly, fairly, and justly. Triage decisions should not be made in an arbitrary or prejudiced fashion. Ethnic origin, race, sex, creed, social worth, sexual preference, and ability to pay should never be factors in determining triage decisions.

Triage policies should be disclosed in advance to the general public and when, feasible, to patients and surrogates on admission. Triage decisions may be made without patient or surrogate consent. Disclosure of triage decisions may help to facilitate communication, understanding, and corporation among patients, surrogates, and physicians.

As a general rule, obligations to patients already hospitalized in an ICU who continue to warrant ICU care outweigh obligations to accept new patients. There may be circumstances, however, when it is justified to discharge a patient from the ICU to admit another patient (see next section). If admission of a new patient is likely to adversely affect the outcomes of patients already under care, then that admission is ordinarily justified only if the benefit to the new admission in significant and quite likely and the adverse effects on the present ICU patients are either conjectural or unlikely to be significant.

Criteria for ICU admission, discharge, and exclusion should be explicitly described. Patients who have do-not-resuscitate orders (no cardiopulmonary resuscitation, no chest compression) may still have compelling reasons for admission to ICU and should not be automatically excluded. Patients with little or not anticipated benefit from further ICU treatment may be discharged or transferred from the ICU. Patients with terminal, irreversible illness who face imminent death should be excluded from ICU. Very elderly individuals who are failing to thrive due to irreversibly, chronic illness should not be encouraged to use intensive car. The decision to exclude or discharge a patient from the ICU may be appropriately be made despite the anticipation of an untoward outcome.

Examples of terminally ill patients who *may* be excluded from the ICU, whether beds are available or not, include those with severe, irreversible brain damage or irreversible multiorgan failure and those with metastatic cancer unresponsive to chemotherapy and/or radiation therapy unless the patients are in specialized ICUs or on specific protocols.

Examples of patients who *should* be excluded from the ICU, whether beds are available or not, include those who competently decline intensive care or request that invasive therapy be withheld; those declared brain dead who are not organ donors[19]; and those in a persistent vegetative or permanently unconscious state.[19]

However, religious or moral convictions may legitimately be basis for the provision of treatment to such patients if the costs are not borne by the general society and the provision of such services does not foreclose the treatment of other patients who would benefit from critical care.[24]

Although triage decisions in military and disaster situations may mandate prioritizing care on the basis of potential contribution to military goals or the saving or other lives, such considerations are not ordinarily applicable to the civilian ICU setting. Hospitals should have a separate policy to handle civilian disasters.

A patient's personal behavior should not influence triage decisions if it does not affect the patient's outcome.

Patients, their surrogates, or others may not compel a physician to provide treatment that the physician believes is not medically indicated.

IV. MECHANICS

Institutional Responsibilities

Hospitals are responsible, acting through ICU directors, for setting policy and implementing triage decisions in ICUs.

This responsibility will ordinarily be implemented by the critical care unit (ICU) committee. Depending on the administrative structure of the hospital, this responsibility may be overseen by other committees and/or individuals. The mechanism chosen will depend on each institution's view of the best method to discharge this responsibility.

Because of the complex medical, ethical, and legal issues involved in ICU triage, the ICU committee should include a broad range of health care professionals and other appropriate individuals. When appropriate, the ICU committee should seek the advice of members of the medical and nursing staffs, hospital administration, hospital ethics committee, patient advocacy representatives, and/or legal advisers.

The ICU committee should have medical oversight and responsibility for triage decisions. In addition, it should be vested with authority to implement these decisions.

The ICU committee should institute a mechanism for postponing elective operations and procedures requiring subsequent ICU care when ICU beds are not available.

The ICU committee should assess outcomes on a regular basis to ensure that policies do not lead to inappropriately restrictive ICU admission policies, premature discharge policies, misuse of ICU facilities, unwarranted morbidity and mortality, inappropriate costs.

In its oversight role regarding triage, the ICU committee has the following responsibilities:

- evaluation of backup capacity, such as the capacity of the postanesthesia care unit or the emergency department to handle an overflow of critically ill patients;
- review of policies related to intermediate care or step-down units and their capacity and function;

- maintenance of a program for the regional cooperation among hospitals during periods of overload;
- oversight of the functioning of the triage officer in relation to ICU admission and discharge policies and the triage of individual patients; and
- education of members of the institutional staff about triage issues and policies through appropriate mechanisms.

The Triage Office

On a daily basis, a single individual should have the authority to implement hospital triage policy.

When all critical care units are filled to capacity, the hospital triage officer, with the cooperation of the ICU directors, should have access to all critical care units and have responsibility and authority to admit patients to and discharge patients from these units. The triage officer has the ultimate responsibility and discretion to implement hospital policy by arranging transfer of patients in and out of ICUs.

If there is an irreconcilable disagreement between the triage officer and other interested parties, the ICU committee should have mechanism for reviewing appeals.

The triage officer should be a senior, experienced, and respected physician. The officer may delegate routine decisions to other physicians and nonphysicians as appropriate. The officer should make decisions in consultation with others who have appropriate expertise. When a physician acts as both health care provider and triage officer, conflicts of interest are inevitable. Physicians serving in both roles must be aware of possible conflicts. Where adequate facilities exist to assign different personnel to each role, institution should do so. When dual assignments are unavoidable, triage officers should balance both responsibilities impartially and should seek the advice of others. The triage officer's decisions may have to be monitored by the ICU committee.

Agreements between physicians and their patients regarding availability of ICU patients regarding availability of ICU care are subject to the oversight and possible modification of the triage officer.

Hospital triage officers should have general knowledge of the various conditions treated in the hospital's ICUs, treatments commonly employed, and their associated range of outcomes. They should also be familiar with the various prognostic models and their uses and limitations.

Regardless of the patient's circumstances, the process of triage has the following elements:

- patient assessment;
- urgency determination;
- priority of care based on urgency;
- resource analysis;
- documentation; and
- disposition.

Triage involves a common logic that includes probability estimates of outcomes; judgements of the benefit and burden to the patient and the system; and judgments of the value of the outcome of the patient and to the system.

Methods of Communication

The hospital should keep those who need to know informed about the status and availability of ICU beds. Physicians whose patients may require ICU services have a responsibility to know hat is available and communicate their patients' needs to the appropriate individual.

Hospital triage officers should communicate effectively with those caregivers affected by triage decisions. The triage officers should engage in active and explicitly negotiation and consultation with other caregivers when making decisions that may include postponing elective procedures for patients who normally require assignment of ICU beds.

REFERENCES

1. Thurow LC. Learning to say no. *N Engl J Med.* 1981; 311:1569–1572.
2. Evans RW. Health care technology and the inevitability of resource allocation and rationing decisions. *JAMA.* 1983; 249:2047–2053, 2208–2219.
3. Aaron HJ, Schwartz WB. *The Painful Prescription: Rationing Hospital Care.* Washington, DC: The Brookings Institution; 1984.
4. Fuchs VR. The 'rationing' of medical care. *N Engl J Med.* 1984; 311:1572–1573.
5. Berenson RA. *Intensive Care Units (ICUs): Clinical Outcomes, Costs and Decisionmaking.* Washington, DC: Office of Technology Assessment; 1984. Health Technology Case study 28, US Congress OTA-HCS–28.
6. Hiatt H. *America's Health in the Balance: Choice or Chance?* New York, NY: Harper & Row Publishers Inc; 1987.

7. Churchill LR. *Rationing Health Care in America: Perceptions and Principles of Justice*. Notre Dame, Ind: Notre Dame Press; 1987.
8. Schneider EL, Guralnik, JM. The aging of America: impact on health care costs. *JAMA*. 1990; 263:2335–2340.
9. Strauss MJ, LoGerfo JP, Yeltatzie JA, Temkin N, Hudson LD. Rationing of intensive care unit services: an everyday occurrence. *JAMA*. 1986; 255:1143–1146.
10. Engelhardt HT, Rie MA. Intensive care units, scarce resources, and conflicting principles of justice. *JAMA*. 1986; 255:1159–1164.
11. Kalb PE, Miller DH. Utilization strategies for intensive care units. *JAMA*. 1989; 261:2389–2395.
12. Rund Da, Rausch TS. *Triage*. St. Louis, Mo: CV Mosby Co; 1981.
13. Winslow GR. *Triage and Justice*. Berkeley, University of California Press; 1982.
14. Reagan MD. Health care rationing: what does it mean? *N Engl J Med*. 1984; 311:1572–1573.
15. Baker R, Strosberg M. Triage and equality: an historical reassessment of utilitarian analyses of triage. *Kennedy Inst Ethics J*. 1992; 2:103–123.
16. Teres D. Civilian triage in the intensive care unit: the ritual of the last bed. *Crit Care Med*. 1993; 21:598–606.
17. President's Commission for the Study of Ethical Problems in Medicine and Behavioral Research. *Deciding to Forego Life-Sustaining Treatment: A Report on the Ethical, Medical and Legal Issues in Treatment Decisions*. Washington, DC: US Government Printing Office; 1983:3.
18. Danis M, Churchill LR: Autonomy and the common weal. *Hastings Cent Rep*. 1991; 21:25–31.
19. Task Force on Guidelines, Society of Critical Care Medicine. Recommendations for intensive care unit admission and discharge criteria. *Crit Care Med*. 1988; 16:807–808.
20. Critical care medicine. *JAMA*. 1983; 250:798–804. NIH Consensus Conference.
21. Lantos JD, Singer PA, Walker RM, et al. The illusion of futility in clinical practice. *Am J Med*. 1989; 87:81–84.
22. Schneiderman LJ, Jecker NS, Jonsen AR: Medical futility: its meanings and ethical implications. *Am Intern Med*. 190; 112:949–954.
23. Truog RD, Brett As, Fader J: The problem with futility. *N Eng J Med*. 1992; 326:1560–1564.
24. Olik RS: Brain death, religious freedom, and public policy: New Jersey's landmark legislative initiative. *Kennedy Inst Ethics J* 1991; 1:275–288.

Originally Published
The Journal of the American Medical Assocation
April 20, 1994, Vol. 271, No. 15
Reprinted with Permission

NOTES ON CONTRIBUTORS

Baruch A. Brody, Ph.D., Director, Center for Ethics, Medicine, and Public Issues, Baylor College of Medicine, Houston, Texas, 77030, U.S.A.

Herman H. Delooz, M.D., Ph.D., F.C.C.M., Director, Department of Emergency Medicine, University Hospitals, Catholic University of Leuven, Leuven, Belgium

Diego Gracia, Ph.D., Department of the History of Medicine, Ciudad Universitaria, 28040 Madrid, Spain

M. Cathleen Kaveny, J.D., Ph.D., Ropes & Gray, 1 International Place, Boston, Massachusetts, U.S.A.

Thomas R. Kopfensteiner, Ph.D., Kenrick School of Theology, 5200 Glennon Drive, St. Louis, Missouri, 63119–4399, U.S.A.

Michael A. Rie, M.D., F.A.C.P., Department of Anesthesiology, A.B. Chandler Medical Center, University of Kentucky, Lexington, Kentucky 30536–0084

Pedro Rosso, M.D. , School of Medicine, Pontifical Catholic University of Chile, Casilla 114-D, Santiago, Chile

Miguel A. Sanchez-Gonzalez, M.D., Ph.D. Universidad Complutense, Madrid, Spain

Paul T. Schotsmans, Ph.D., Director, Centrum Voor Bio-Medische Ethiek, Catholic University of Leuven, Leuven, Belgium

Alejandro Serani-Merlo, Ph.D. Center for Bioethics, School of Medicine,Pontifical Catholic University of Chile, Casilla 114-D, Santiago, Chile

Kevin Wildes, S.J. Ph.D., Department of Philosophy and The Kennedy Institute of Ethics, Georgetown University, Washington, D.C., 20057, U.S.A.

Index

K. Wm. Wildes (ed.), Critical Choices and Critical Care, 217–220.